Your Money, Their Algorithm

How AI Is Changing Investing, Insurance, and Your Financial Life

Joseph Trek

Published by NobleTrex Press

For permissions and other inquiries, write to:

P.O. Box 3132, Framingham, MA 01701, USA

Contents

Introduction

It takes approximately two hundred milliseconds for the blinking cursor on your screen to confirm your loan application, execute your stock trade, or quote your car insurance premium. That is roughly half the time it takes you to blink your eyes. In that imperceptible sliver of time, a digital jury has convened, reviewed the evidence of your financial existence, deliberated, and delivered a verdict that will shape your economic reality.

They didn't just look at your bank balance. They looked at how fast you typed your address. They checked if you capitalized your last name. They cross-referenced the geolocation of your phone with the merchant code of the coffee shop where you're sitting. They queried a database of people who share your purchasing habits to predict the likelihood that you will default on a debt three years from now.

Before your thumb even left the glass of your smartphone, the decision was made. You were quantified, categorized, and priced.

For most of history, money was a relationship. You knew the banker; the banker knew your parents. Risk was a gut check made by a human being across a mahogany desk. It was slow, it was biased, and it was often unfair. Today, money is a data stream. The relationship has been replaced by a prediction engine-an intricate web of algorithms designed to maximize efficiency and extract value. The mahogany desk is gone, replaced by a server farm in Northern Virginia that hums twenty-four hours a day, processing the chaotic inputs of your life into neat, tradable probabilities.

We have traded the awkwardness of human judgment for the ruthless efficiency of the black box. On the surface, this trade looks like a bargain. We have commission-free trading, instant approvals, and apps that nudge us to save for a rainy day. But beneath the sleek user interfaces and gamified progress bars lies a new set of rules that most of us never signed up for. We are playing a game where the referee is a robot, the rulebook is written in proprietary code, and the house-as always-has calculated the edge down to the tenth decimal point.

This shift is not inherently bad. In fact, it has democratized wealth in ways that were unimaginable a generation ago. A twenty-year-old with a smartphone today has access to investment tools that hedge fund managers in the 1990s would have killed for. But it has also introduced a dangerous new form of illiteracy. We know how to use the apps, but we don't understand the machinery beneath them. We are passengers in a self-driving car, napping in the backseat, assuming the computer sees the cliff edge that is approaching.

You picked up this book because you suspect that the "smart" recommendations you receive aren't always for your benefit. You picked it up because you want to know why your credit limit dropped, why your insurance rates spiked despite a clean record, or why your portfolio seems to churn even when you haven't touched it. Or perhaps you simply want to know if the robots are going to make you rich or retire on your dime.

The answer is complex, fascinating, and urgent. We are living through the greatest rewiring of the financial system since the invention of the ledger. To survive it-and to thrive in it-you need to understand how the machine thinks.

Let's be honest about how we currently manage money. We treat our financial apps like magic 8-balls. We shake the phone, ask for guidance, and accept the result. We employ "robo-advisors" to manage our retirement, assuming that an algorithm is inherently objective. We trust that if an app is free, it's a public service, ignoring the colossal data economy that keeps the lights on.

The first thing we need to dismantle is the myth of the neutral algorithm. There is no such thing as "math without an opinion." Every line of code in your banking app was written by a human with a specific incentive structure. When a platform suggests a portfolio allocation, is it optimizing for your long-term tax liability, or is it optimizing for the platform's need to route orders to a specific market maker? When a lender's AI approves your credit card, is it betting on your financial success, or is it predicting that you are the perfect, profitable customer who will carry a balance but never quite default?

We often confuse automation with benevolence. We assume that because a computer is doing the math, the answer must be the absolute truth. But in the world of algorithmic finance, the answer is often just the most profitable probability. By understanding the incentives of the platforms we use, we stop being the product and start being the client again. We have to learn to look at the "free" tools in our pocket and ask: *Who is paying for this, and what are they buying?*

But this isn't just about skepticism; it's about mechanics. The digitization of money has changed the physics of the markets. Stocks don't move the way they used to. We live in an era where high-frequency trading firms fight over microseconds, and where retail sentiment can be weaponized by algorithms scraping social media for keywords. The price you see on your screen is not a static number; it is a fleeting consensus between warring mathematical models. If you don't understand how your order gets executed, you are paying a tax you can't see on every single transaction.

Furthermore, the algorithm has left the stock market and entered your driveway. The insurance industry is undergoing a quiet revolution that will arguably impact your wallet more than any stock market crash. The old model of "pooled risk"-where the safe drivers subsidized the risky ones-is dying. It is being replaced by surveillance pricing. Your car, your watch, and your thermostat are talking to your insurer. They are building a digital twin of your behaviors. This promises lower rates for the "perfect" citizen, but it raises profound questions about privacy and the purpose of insurance. If the algorithm knows you are going to get sick or crash your car before you do, is it still insurance, or is it just a prepayment plan for inevitable disaster?

Despite the futuristic landscape, there is a comforting irony at the heart of this revolution: The smarter the machines get, the more the old, boring laws of wealth matter.

Silicon Valley loves to disrupt, but it cannot disrupt arithmetic. No amount of machine learning can change the fact that spending more than you earn leads to ruin. No artificial intelligence can repeal the laws of compound interest or the relationship between risk and return. In fact, the barrage of data and the speed of execution make the ancient principles of patience, diversification, and liquidity more valuable, not less.

The danger is that the technology is designed to make us abandon these principles. The interface of modern finance is designed to trigger the dopamine receptors in your brain, not the prudent planning centers of your prefrontal cortex. When you deposit money, confetti explodes on the screen. When stocks move, they flash red and green like a slot machine. Push notifications create a sense of urgency that is antithetical to long-term investing. The algorithm knows that an active, anxious user is a profitable user. It knows that if you check your balance ten times a day, you are more likely to transact. It knows your biases better than you do-it knows you sell when you're scared and buy when you're greedy, and it is ready to take the other side of that trade.

We are going to explore how to build a firewall against this psychological hacking. We will look at how to use the incredible power of automation to enforce good habits rather than bad ones. We can rig the game in our favor, setting up digital guardrails that save us from our own worst impulses.

This is the great promise of the algorithmic age. If we can harness these tools properly, we can remove the friction from wealth building. We can tax-loss harvest instantly, rebalance portfolios while we sleep, and detect fraud before the charge clears. We can use the machine to handle the execution, leaving our minds free to focus on the strategy.

But what about the future? We are standing on the precipice of the Generative AI era. The chatbots are already here, offering financial advice that sounds confidently authoritative, even when it is disastrously wrong. We are moving toward a world where you might have a personalized AI CFO in your pocket-an agent that negotiates your cable bill, optimizes your credit card rewards, and tells you exactly how much you can spend on dinner tonight without jeopardizing your retirement.

It sounds like utopia, but it requires a new kind of vigilance. When you outsource your financial thinking to a Large Language Model, you are trusting it not just with your money, but with your definition of "enough." We have to learn where the AI hallucinates, where it oversimplifies, and where it fails to understand the nuance of human life. A retirement plan is not just a math problem; it is a statement of values. An algorithm can calculate the most efficient drawdown rate for your 401(k), but it cannot calculate the value of passing a family business to a child or the peace of mind that comes from a paid-off mortgage.

We must remain the pilots. The technology is the navigation system, the autopilot, and the engine management software. It is brilliant, tireless, and fast. But it doesn't know where we want to go.

In the chapters that follow, we are going to tear down the sleek facade of modern finance and look at the wiring underneath. We will start by auditing your digital existence, understanding exactly what data you are broadcasting to the world and how it is monetized. We will demystify the "black box," explaining in plain English-not PhD statistics-how machine learning models actually make predictions and why they crash.

We will walk through the mechanics of the robo-advisor, dissecting how a questionnaire becomes a portfolio and where the hidden fees are buried. We will look at the invisible auction of the credit markets and the surveillance state of modern insurance. We will tackle the uncomfortable reality of "algorithmic bias," where neutral code produces discriminatory outcomes, and what you can do to protect your score.

We will also look at the unchanging laws of money-the bedrock principles that will save you when the algorithm glitches. We will analyze the behavioral tricks used by trading apps to turn investors into gamblers, and we will rewrite the notification settings to serve your sanity rather than their engagement metrics.

Finally, we will construct a governance framework for your life. You are the CEO of your household finances. The apps, the bots, and the algorithms are your employees. They are powerful, efficient workers, but they require supervision. You cannot simply set them and forget them; you must understand them, monitor them, and know when to override them.

The era of the handshake deal is over. The era of the algorithm is here. It is faster, sharper, and more complex than anything that came before. It can be a terrifying master, capable of locking you out of the economy in milliseconds. But if you understand how it works, if you respect the math and retain your agency, it can be the most powerful servant you have ever had.

Let's look under the hood.

Chapter 1

The New Rules of Money

Money used to be physical-coins in a purse, ledgers in a safe. Today, money is data. Before we can understand how AI algorithms trade stocks or assess insurance risks, we must understand the digital plumbing that carries your wealth from point A to point B. In this chapter, we strip away the slick user interfaces of your favorite fintech apps to reveal the new rules of the game: how your cash flow creates a digital fingerprint, why 'free' tools often come with a hidden price tag, and how to build a financial foundation that remains solid even when the software glitches.

1.1 Cash Flow and the Digital Balance Sheet

At 11:47 p.m., your phone lights up with a congratulatory push notification: *Nice work—you can afford to save $312 this week. Want us to move it now?* It feels like praise. It's also a proposal. Somewhere behind that cheerful sentence, a model has taken your life—rent, groceries, late-night takeout, the one oddly expensive Tuesday at the pharmacy—and compressed it into a prediction about tomorrow morning's balance. The app isn't admiring your discipline. It's reading your pattern.

For most of financial history, money was measured in snapshots. A bank statement. A credit score. A net worth calculation you did once a year and immediately regretted. Algorithms prefer something else: motion. They want to know not only what you *have*, but what you *do*. The modern financial system increasingly treats your personal finances like a stream of telemetry, and two old-school ideas—net worth and cash flow—have become the most important columns in the dataset.

Net worth is the simplest: everything you own minus everything you owe. It's your financial *mass*. If you sold your investments, paid off your debts, and closed the tabs in your brain that always keep a running total, net worth is what would remain. Cash flow is different. Cash flow is your financial *velocity*: money moving in and out, day after day. Wages landing. Bills leaving. Subscriptions nibbling. Refunds bouncing back. Tips, transfers, reimbursements, surprises. Net worth tells a story about where you stand; cash flow tells a story about how you live.

If you've ever looked "wealthy" on paper while feeling broke, you've already met the gap between mass and velocity. A friend with a high salary and no savings can have impressive cash flow and fragile net worth. A retiree with a paid-off home and modest income can have strong net worth and tight cash flow. Neither picture is wrong. But they lead to different outcomes when the decision-maker is not a person with context, but software with rules.

This is where the idea of a *digital balance sheet* comes in. Your digital balance sheet isn't just the spreadsheet you make for yourself. It's the version of your finances that gets inferred and reconstructed by financial systems—sometimes from information you explicitly provide, sometimes from data they receive because you linked accounts, used a card, or clicked "agree." It's a practical concept, not a conspiracy. Lenders have always built internal views of a customer. What changed is the *resolution*. They no longer need a handful of numbers; they can ingest the whole rhythm.

In that rhythm, consistency is a kind of power. Algorithms love regularity the way engineers love clean signals. A paycheck that arrives every other Friday is easy to model. Rent on the first is easy to anticipate. A stable grocery budget makes you look predictable. Predictability reduces uncertainty, and uncertainty is what risk models are paid to squeeze. This is one reason two households with the same income can be treated differently: one looks like a smooth waveform, the other like a seismograph.

The seismograph household is not irresponsible. It might be a freelancer with lumpy invoices, a nurse with variable shifts, a family managing seasonal expenses, or a student whose income comes in bursts. The problem is not the reality; it's the *representation*. A model that was trained on clean, salaried patterns can misread irregular-but-healthy behavior as instability. In a human conversation, you'd say, "My income is uneven, but it averages out." In a dataset, "averages out" can vanish under the noise of timing.

Here's a small, ordinary example that reveals how digital balance sheets are made. Imagine you get paid on the last business day of the month, but rent is due on the first calendar day. On paper, you are fine. In practice, there's a recurring two-day period where your account is low. A person understands the calendar; an automated system might see "frequent low balance events." If your app runs an auto-save feature that pulls money out when it thinks you have "extra," that two-day gap can turn into an overdraft fee. The model doesn't hate you. It's just optimizing on a simplified view of time.

Net worth, meanwhile, has its own algorithmic pitfalls. You can have a high net worth that is not *liquid*—meaning it can't easily be turned into cash without friction. Home equity is real wealth, but it doesn't pay the electric bill on Tuesday. Retirement accounts are real wealth, but early withdrawals can be expensive. A person feels the difference between "I'm worth $500,000" and "I have $500." An app can blur it if it doesn't sharply separate liquid assets from long-term ones, or if it treats "available" as "accessible."

That distinction—liquid versus illiquid—is one of the first upgrades you can make to your own financial thinking, because it's also one of the first upgrades you can make to how machines treat you. Net worth is not one number; it's a stack of categories with different rules. Cash in checking is not the same as cash in a brokerage sweep account. A brokerage account is not the same as a 401(k). A paid-off car is not the same as a paid-off house. In a digital balance sheet, mislabeled categories can produce very real consequences: the wrong "safe-to-withdraw" assumption, the wrong "emergency fund" suggestion, the wrong "you can afford this" nudge.

The story gets more interesting when you realize that cash flow is not merely a budgeting concept anymore; it has become a risk signal. A traditional lender might have asked for pay stubs and glanced at your debt-to-income ratio. A cash-flow-informed lender can look at the last 12 months of deposits and withdrawals and ask different questions: How often does your balance hit zero? How many overdraft fees appear? Do you pay bills on time? Is your income concentrated in one source or spread across several? Do you receive irregular deposits that look like gig work? Models can learn these patterns even when no one is "reading" your transactions line by line.

That can cut both ways. For someone with a thin credit file—a young adult, a recent immigrant, a person who avoided credit cards—cash flow data can provide a richer picture than a score that mostly measures how long you've borrowed. For someone whose life includes medical surprises, caretaking costs, or variable income, that same richness can become a microscope. The digital balance sheet doesn't judge your story. It quantifies your volatility.

Volatility is the word investors use for price swings, but it belongs in personal finance too. A month where you spend \$3,000 is not a problem if you normally spend \$3,000. A month where you spend \$3,000 when you normally spend \$1,800 is a spike, and spikes are what algorithms treat as anomalies. The modern "smart" financial experience often works by detecting anomalies: fraud detection, budgeting categories, subscription cancellation suggestions, debt payoff plans, auto-invest, auto-save. An anomaly can be a stolen card—or it can be a wedding gift you bought on a credit card because you were short on cash that week. If the system can't tell the difference, it might "help" at the wrong moment.

The way out is not to avoid automation. It's to automate *on purpose*. The most common failure mode in personal fintech is letting a tool decide what your goal is. Tools are enthusiastic.

They're always offering to optimize something. The problem is that optimization is only as good as the target. If you tell an app, explicitly or implicitly, that the target is "maximize savings," it may happily starve your checking account. If you tell an investment platform the target is "maximize returns," it may lead you into risk you can't emotionally tolerate. Machines don't understand the pain of an overdraft fee or the shame of a declined card. They understand a constraint only if you express it as a rule.

That's why the best financial goals are machine-readable. Not because you're trying to live like a robot, but because you're trying to keep the robot from living *instead* of you. "Save more" is a human wish. "Move $150 every payday into a separate account, but never let checking drop below $800" is a rule. "Invest aggressively" is a vibe. "Contribute $300 per month to a total market index fund and stop contributions automatically if I lose my job" is an executable plan.

Notice what makes the rule work: it includes a *trigger* (payday), an *amount* ($150), a *destination* (separate account), and a *floor* ($800). The floor is the part most people forget. Floors and ceilings are how you tell a system what "safety" means. Without a floor, "smart saving" tools can become financially rude guests: always taking a little more because they assume the party won't end. With a floor, you've drawn a boundary the model can respect.

You can do the same for debt. "Get out of debt" is emotionally true and operationally vague. A machine-friendly version sounds more like: "Pay the minimum on all debts, then send $200 extra to the highest-interest card each month, but pause extra payments during months when my cash flow is negative." That last clause is not moral weakness; it's antifragility. It prevents the system from chasing an ideal plan at the expense of your ability to pay rent.

A digital balance sheet also includes your *timing*. Timing is the hidden dimension most budgets ignore, and it's where a lot of automation breaks. You can be "within budget" and still run out of cash mid-month if bills are front-loaded and income is back-loaded. Algorithms that forecast "end-of-month balance" can miss the fact that you have a low-balance valley on the 12th. If you want your tools to work, you have to give your money a calendar, not just categories.

One way to do that is to think in terms of *mini-balance sheets* within the month. Your paycheck lands; you immediately reserve money for fixed bills; you set aside a buffer; then you allow flexible spending. Humans often do this mentally ("I can't touch that, it's for rent"). Apps sometimes do it poorly ("We'll move money around as we see fit"). The trick is to make the mental structure explicit so automation doesn't bulldoze it.

There's also a subtle social dimension. A lot of cash-flow irregularity comes from being a helpful person: fronting money for a group dinner, paying for kids' activities and getting reimbursed later, sending money to family, covering a friend in a pinch. Your bank account sees the outflow; the reimbursement might land days later through a different app. To you, it's neutral. To a model, it can look like spending spikes followed by random deposits. If you rely heavily on reimbursements, your digital balance sheet can misclassify your behavior as instability unless you corral it.

"Corral it" can be as simple as one dedicated account for reimbursements, or one payment method for group expenses, so the pattern is legible. Machines are not wise, but they are good at pattern recognition *if* you give them consistent signals. The goal is not to perform a tidy life. It's to reduce accidental ambiguity in the data trail you're already generating.

Another blind spot: subscriptions. Subscriptions are the perfect machine food: consistent, labeled, predictable. They're also quietly corrosive because they create the illusion of a fixed lifestyle while slowly raising your baseline expenses. Algorithms that suggest "safe to save" amounts often assume your recent baseline spending is a stable truth. If subscriptions creep upward, the model updates its view of "normal" and starts saving less—not because it's wrong, but because it's obedient. In other words, your financial life can get more expensive without any single moment of decision, and your "smart" tools may calmly accept the new normal.

Net worth has a creep problem too, but in reverse: invisible liabilities. Buy-now-pay-later plans, installment offers, and "split it into four" checkouts can keep purchases feeling small while building a shadow schedule of future obligations. You may not experience them as "debt" because they don't resemble a credit card. But cash flow experiences them as very real. When many small obligations stack up, you can have a month that looks fine—until the week when three installments and a utility bill collide. Digital balance sheets care about collision.

The unifying idea is that personal finance is no longer just self-knowledge; it's also *system behavior*. Your money habits are signals. Some signals are deliberate (you set a savings transfer).

Others are incidental (you get paid at 2 a.m. because payroll processed early). Financial algorithms act on both. The art is to turn more of your signals into deliberate ones.

A practical framework: think of your finances as two dials and a safety catch. The first dial is *mass*: net worth, built slowly by investing, paying down principal, and accumulating assets. The second dial is *velocity*: cash flow, managed daily by timing, buffers, and spending habits. The safety catch is *liquidity*: cash you can access without selling assets or borrowing at a bad moment. Many financial mistakes happen when you try to turn the mass dial (long-term wealth) using velocity tools (short-term cash), or when you try to fix velocity problems by pretending mass is liquid.

The emotional trap is that net worth feels like identity. It's a scorecard you can obsess over. Cash flow feels like chores. But cash flow is the part that decides whether your life is calm. Algorithms know this, which is why they keep inserting themselves into cash flow: "round-ups," "safe-to-spend," "smart transfers," "instant cash advances," "automatic bill negotiation." If you don't specify boundaries, the system will happily optimize for engagement, for product adoption, or for a simplistic definition of "health."

A good boundary starts with naming your non-negotiables in numbers. What is the smallest balance that lets you sleep? What is the minimum buffer that prevents overdrafts even if a bill posts early? How many days of expenses do you want in immediately accessible cash, not because someone told you a rule, but because your cash flow is lumpy? The answers vary wildly. The point is that you are turning "I hate surprises" into a line of code.

Then you make your finances legible to your own tools. That sounds abstract until you do it once. A separate "bills" account can turn a chaotic month into a predictable system. A scheduled transfer that happens the day after payday can prevent accidental overspending without any willpower. A single category for "true irregulars" (car repairs, gifts, annual fees) can stop you from treating predictable-but-infrequent expenses as emergencies. This is not about perfection; it's about reducing the gap between what you mean and what your data implies.

One more counterintuitive truth: sometimes the best way to improve your digital balance sheet is to *stop* trying to look good and start trying to look clear. People often hide cash flow messiness by moving money constantly—between accounts, apps, cards, and payment services—until nothing is understandable. But constant shuffling can make your financial life harder to model for both you and your tools. Clarity usually comes from fewer accounts doing more specific jobs, not more accounts doing vaguely overlapping jobs.

Your money will never be fully captured by a dataset, and you shouldn't want it to be. Love, stress, ambition, exhaustion—none of that fits neatly into transactions. But the parts that *do* fit into transactions are increasingly influential, because they're the parts machines can act on at scale. The digital balance sheet is the version of you that gets to vote in automated decisions.

Actionable takeaway: Pick one automated feature you use (auto-save, auto-pay, round-ups, auto-invest) and add a hard safety rule to it within the next 10 minutes: a minimum checking balance floor, a monthly cap, or a pause condition tied to income. If the tool can't express a boundary you care about, that's not a personal failure—it's a product limitation, and it's a sign to simplify or switch before "helpful" becomes expensive.

1.2 Banking and Payment Rails: The Plumbing

The weirdest part of modern money is how confidently it lies to your face.

You tap your phone at a coffee shop and your screen says *Approved*. The barista smiles. You walk out holding caffeine and certainty. But under the glassy interface, your payment is more like a promise scribbled on a sticky note: "I will settle up soon." Most of the time, it does. Sometimes it doesn't. And in the gap between *approved* and *settled*, expensive things happen—especially when an app is also trying to "help" by moving your money around.

The pipes matter because algorithms don't wait for plumbing. A budgeting app can classify your purchase instantly. A bank can decide you're "safe to save" in milliseconds. A fraud engine can freeze a card before you've even put it back in your wallet. Yet the underlying money movement—between banks, between institutions, across networks—often runs on schedules, batches, cut-off times, and rules that were built for a world where "instant" meant "same business day."

Most people learn this only when the system misbehaves. A rent payment bounces even though you *swear* you had the money. A paycheck shows as "pending" while autopay quietly pulls the actual dollars out. A transfer you initiated on a Saturday behaves like it's taking the weekend off. You call customer service, and the answer—no matter how politely delivered—boils down to: *the app's screen and the bank's reality are not the same thing*.

Start with a basic picture. When money moves, it usually travels on one of several "rails," which is just a practical term for the network and rules that carry value from one place to another. Cards have rails. Bank-to-bank transfers have rails. Wire payments have rails. Some rails are fast and final; some are slow and reversible; some are "instant" in the user interface but staged behind the scenes. If you want to understand why cash reserves are a form of self-defense, you need to understand what can go wrong inside those rails without anyone being malicious.

Card payments are the most familiar illusion. When you pay with a debit card or credit card, your bank doesn't always send money to the merchant's bank right away. First comes *authorization*: a quick check that you have available credit (or funds) and that the transaction looks legitimate. Your bank may "hold" an amount—reducing what you can spend—before any final settlement happens. Then later comes *clearing and settlement*, where the actual funds move and the details get finalized.

That difference—authorization now, settlement later—is why tips can make restaurant charges change, why hotel deposits feel like they "double-charge" you, and why gas stations sometimes place a larger temporary hold than you actually spend. Your account's *available* balance reacts immediately; your *posted* balance catches up on its own timetable. If you only track one of those numbers, you're living in the part of the story that's optimized for convenience, not accuracy.

ACH, the Automated Clearing House network, is the rail behind a lot of "boring" money movement: paychecks (direct deposit), utility autopays, mortgage payments, many app-to-bank transfers. ACH is powerful precisely because it's not glamorous. It's designed to move huge volumes of payments efficiently, and it often does it in batches. Think of it less like a teleporter and more like a mailroom: files get collected, sorted, and delivered according to a schedule. That schedule has improved over the years, but the basic logic remains: timing matters.

ACH also has a personality trait that surprises people: it can be either a *push* or a *pull*. A push is when you tell your bank, "Send $200 to my landlord." A pull is when you tell a biller, "You're allowed to take $200 from my account." Pulls are convenient until they're not. If a pull hits earlier than expected, or hits twice due to a glitch, you may be right in principle and broke in practice while you dispute and unwind it. In the real world, that's not a philosophy debate; it's groceries.

Then there are wire transfers, like Fedwire in the U.S., which are built for large-value, time-critical payments. Wires are the opposite of ACH in vibe and consequence: fast, direct, and typically treated as final once sent. People don't use wires every day, but their existence is a reminder that "money transfer" isn't one product—it's a menu of different rails with different trade-offs. When something is final, it's fast in the same way a cut is final: you don't "undo" it; you treat it.

Add newer real-time payment systems to the mix and the plot thickens. Real-time networks aim to move money 24/7/365 with immediate confirmation and settlement. That's a big deal for payroll, gig-economy payouts, insurance claims, and any moment where "I need it now" is not drama but math. But even when the rail supports instant movement, the institutions on either end may still wrap it in policies: fraud checks, limits, holds, and review queues. "Real-time" is a capability; your experience is a product decision.

So where does algorithmic error come in? Right where the rails change hands.

A typical modern financial life is a relay race. Your paycheck lands by ACH. Your card swipes create authorizations. Your bills pull money on schedules. Your savings app "optimizes" transfers based on predicted cash flow. Your investment app auto-invests on payday. Your credit card autopay tries to be helpful by paying the statement balance. Each piece might be reasonable in isolation. Together, they can create a chain reaction—because each tool is operating on its own version of "truth," and those truths are updated at different speeds.

Picture a Thursday afternoon. Your checking account has $1,400. Your rent is $1,200 and will pull tomorrow. Your savings app sees a pattern: you "usually" keep about $800 after bills. It decides that $300 is safe to move into savings today. Meanwhile, your card has a pending hotel hold of $250 from a work trip, and your utility company posts a bill a day early. None of these events is outrageous. But you've just learned the difference between "safe according to a model" and "safe in the plumbing."

The failure mode is often timing plus confidence. Algorithms tend to be *decisive*. They don't hesitate because hesitation is hard to quantify. If the model thinks it's 92% sure you won't need $300, it will happily move $300. Humans, when they're 92% sure, usually keep a little slack anyway. That slack is what liquidity is: not extra money for a future vacation, but extra money for the system's small humiliations—holds, reversals, delays, duplicated charges, mis-timed pulls.

Liquidity is often framed as a character trait: disciplined people have emergency funds; sloppy people don't. That moral framing is outdated. Liquidity is an engineering control. You are building a buffer between you and a system that sometimes acts like it's possessed—not because anyone is evil, but because many independent systems are coordinating imperfectly in real time.

There's also the human side of "independent systems." Banks place holds to prevent fraud and losses. Merchants batch transactions to reduce costs. Billers pull early because they can. Payment networks have cut-off times. Customer-service reversals take time because compliance and audit trails are real. Even your employer's payroll department has a calendar. None of these parties is waking up thinking about your rent. They're thinking about their own risk, their own efficiency, and their own rules. Your job is to avoid being crushed by the sum of those reasonable decisions.

A quick detour into the psychology of the app screen: most interfaces show you a single number and call it a balance. That's like showing the weather as a single temperature and refusing to mention wind. In reality, you have several balances at once. There's the posted balance. The available balance. The pending transactions. The scheduled transactions you authorized but haven't hit yet. The autopays you forgot you set up. The transfers that are "in progress." The money that is in your account but held. The money that is not in your account but effectively already spent. When an algorithm is making moves, it is choosing which of these realities to treat as real.

Even the words "pending" and "processing" can hide big differences. A pending card transaction may fall off, post later, or post in a different amount (tips, holds). An ACH transfer may be initiated but not yet settled. A bank-to-bank transfer may show as deducted on one side and not yet credited on the other. The interface turns all of this into a calming progress bar. But money doesn't move on vibes; it moves on settlement.

This is why maintaining liquid cash reserves is not just good advice—it's a defense against automation. If a tool over-saves, your buffer keeps your lights on while you unwind it. If a bill hits early, your buffer absorbs the timing mismatch. If a merchant hold inflates your available balance illusion, your buffer keeps you from stacking transactions on a cliff edge. The buffer doesn't prevent the error; it prevents the error from becoming a crisis.

The most common objection is immediate and fair: "I can't afford a buffer." Often what people mean is, "I can't afford to park money doing nothing." That's a real trade-off. But a liquidity buffer isn't "doing nothing." It's purchasing reliability. It's like paying for good brakes. You don't buy brakes to go faster; you buy brakes so that going fast doesn't kill you.

Reliability has a price, and the system charges it either way. You can pay it upfront by holding some cash, or you can pay it later through overdraft fees, late fees, returned payment fees, payday advances, credit card interest, and the time-tax of calling customer support. When cash is tight, those later costs are especially brutal because they arrive precisely when you're least able to absorb them.

A practical way to think about liquidity is to separate "emergency" from "plumbing." Emergencies are job loss, medical events, family crises. Plumbing is: a charge posted twice; a transfer delayed; a paycheck arriving a day late because of a holiday; a debit card compromised and temporarily frozen; a biller pulling a week early because the due date fell on a weekend. Plumbing problems happen to almost everyone. They don't make headlines. They just quietly drain people who are already trying.

The goal, then, isn't to build a heroic emergency fund overnight. The goal is to build a *settlement buffer*: a cash amount that exists specifically to survive timing mismatch and automated mistakes. It can be smaller than the "six months of expenses" rule-of-thumb and still be life-changing. The settlement buffer is the money that lets you say, "Annoying," instead of, "Catastrophic."

How big should it be? The boring answer is "it depends." The useful answer is: big enough to cover your most dangerous overlap. Your most dangerous overlap is the moment in a typical month when several obligations cluster before income replenishes—rent plus a card payment plus childcare plus utilities, for example. If you've ever had a week where you were juggling dates like a short-order cook, you already know your overlap.

Once you see overlap, you can design around it. Some people solve it with timing: move bill due dates where possible, schedule transfers the day after payday, split rent across paychecks (with a landlord who allows it), pay credit cards more frequently to keep utilization and cash flow smoother. Some solve it with structure: one account for bills, one for spending, one for savings. That's not trendy "bucket" advice; it's operational hygiene. You're creating compartments so one error doesn't flood the whole house.

Automation can be part of the solution if you treat it like a sharp tool. Auto-pay is wonderful until it's set to "statement balance" on a month when your expenses spike. Auto-save is wonderful until it competes with rent. Auto-invest is wonderful until it pulls money that was earmarked for the dentist. The fix is not to turn automation off; it's to install guardrails that reflect settlement reality: minimum checking balance floors, maximum transfer caps, and scheduling that respects your overlap.

There's also a social truth that matters: people with stable income and high balances rarely notice these rails. The plumbing works best for those who can afford to be indifferent to it. If you're living closer to the edge, you become an unwilling expert in bank cut-off times and "when does this actually post?" questions. That's not because you're worse at money. It's because the cost of the system's delays is regressive. A $35 overdraft fee is a rounding error to one household and a grocery budget to another.

And when AI enters the picture, the regressivity can get sharper. Models trained on "typical" cash flow patterns may treat low balances as risk, which can lead to more friction: holds, limits, denials, higher fees, nudges toward expensive credit products. If you don't have a buffer, you may be forced into the very products that extract more from you—cash advances, overdraft lines, high-interest revolving debt—because the system's plumbing mistake created a shortfall. Liquidity is also a way of refusing that funnel.

None of this requires cynicism. It just requires respecting the system as it actually is. The money layer beneath your apps is a patchwork of rails, each with its own timing and rules. Your apps are fast because software is fast. Your money is slower because settlement is cautious. If you plan as if everything is instant, you're planning with the interface, not with reality.

Actionable takeaway: Create a "settlement buffer" in checking (or a linked savings account that you can access immediately) that is *explicitly* off-limits to automation. Give it a number, not a feeling: start with $200 if that's what you can do, or one week of your non-negotiable bills if you can do more. Then set every auto-transfer and auto-payment tool you use to respect that buffer—either by adding a minimum balance floor, capping automated withdrawals, or scheduling automation for the day after income reliably posts. This one move turns payment-rail delays and algorithmic overconfidence from emergencies into inconveniences.

1.3 The Data Economy Beneath Your App

The most consequential button in consumer finance might be *Connect your bank.*

It's never framed that way, of course. The button sits there like a convenience snack: small, friendly, apparently harmless. Tap it and the app suddenly looks brilliant. Your spending categories appear as if by magic. Your paycheck is detected without you typing a single digit. Your "safe-to-save" number updates itself. It feels like a personal assistant moving into your phone.

What actually moved in was a pipeline.

When you link a bank account to an app, you're not just logging in. You're authorizing a data flow-sometimes directly to the app, often through an intermediary whose entire business is being the translator between your bank and everyone else's software. This is the data economy beneath your app: a quiet, always-on set of connections that turns your financial life into something that can be read, copied, scored, compared, and—depending on the permissions—acted upon.

The twist is that the trade-off isn't "privacy versus nothing." The trade-off is privacy versus *automation*. The same data that makes a budgeting app useful also makes it invasive. The same permissions that let a tool warn you about overspending can also let it infer when you got a raise, when you traveled, which pharmacy you use, whether you're supporting someone else financially, and how close you skate to zero between paychecks. This isn't about paranoia; it's about acknowledging that transaction histories are autobiographies written in receipts.

A transaction record is a tiny rectangle of truth: date, amount, merchant, category (sometimes), and a few cryptic words that a bank calls a description. One record is boring. A year of them is intimate. It reveals the shape of your weeks, the stress in your months, the moments you tried to be healthy, the months you didn't. It can tell a story about childcare, medical needs, job changes, breakups, hobbies, addiction, religion, politics, and family obligations-without ever using those words. Money is an information-rich substance, like DNA, except you hand it over dozens of times a day.

Open Banking is the umbrella term for the idea that you should be able to share your financial data-securely, with permission-with third-party services that you choose. Done well, it's pro-consumer. It can make it easier to switch banks, compare products, get better service, and build tools that a single bank might never bother to build. Done badly, it becomes an industry of "consent theater," where you click *I agree* and discover later that you agreed to far more than you imagined.

APIs-Application Programming Interfaces-are the pipes and valves that make Open Banking possible. An API is a structured way for software systems to ask for data and receive it in a standardized format. Instead of a human logging in and clicking around, software can request "account balances" or "transactions for the last 90 days" and get exactly that, formatted cleanly, ready for an algorithm to digest.

If that sounds abstract, here's the real-world feel: APIs are what make it possible for an app to "see" your bank account without you manually exporting a spreadsheet every month. And because they're designed for machines, they can be fast, frequent, and scalable. That's the promise. That's also the risk: once the pipes exist, more things can flow through them than you would ever type into a form.

In the earlier era of fintech, many apps relied on something called *screen scraping*. You would enter your bank username and password into a third party's interface (already a stomach-turning sentence), and that third party would log in on your behalf, pretending to be you, then "scrape" the data from the webpages your bank would normally show you. It worked. It was also fragile, messy, and hard to govern. If your bank redesigned its website, the pipeline could break. If your bank flagged the login as suspicious, you could get locked out. If the third party stored credentials insecurely, the consequences were obvious and ugly.

API-based connections are meant to be better than that. Instead of handing your password to a middleman, you authenticate with your bank and grant specific permissions-often via a token that can be revoked. That's the ideal: scoped access, clearer controls, less credential sharing, fewer brittle workarounds.

But even in the "better" world, the middleman often remains.

Many apps don't connect to thousands of banks individually. They connect to a data aggregator-an infrastructure company that already has the bank connections, the compliance posture, the normalization logic, and the incentive to make the pipe feel seamless. The aggregator becomes the universal adapter for your financial life: one connection point that can feed budgeting apps, lenders, payroll apps, investing dashboards, bill negotiators, subscription trackers, and identity verification services.

Here's what that middleman model looks like in its simplest form:

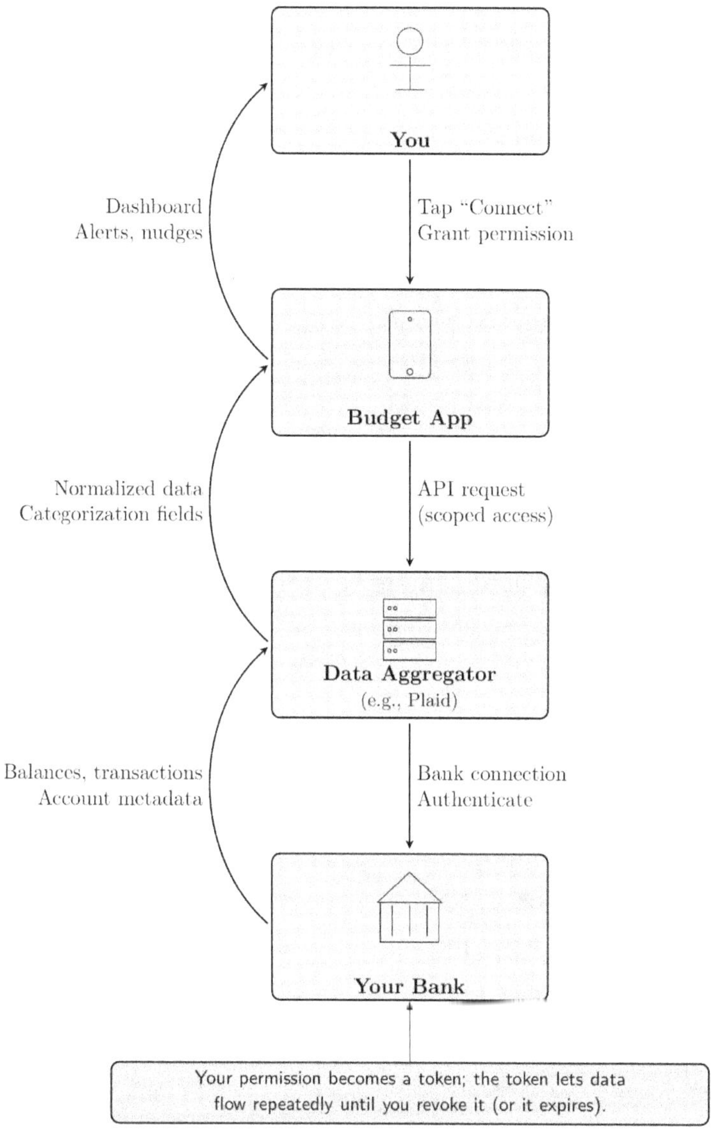

Once you can see the chain, you can start asking better questions. When an app says it "connects to your bank," which connection is it actually using? Is it a direct connection? Is an aggregator involved? Who stores the token? Who decides what "transaction description" fields get passed along? Who is responsible when the connection breaks and the app starts guessing?

The phrase "your data" sounds like a locked box with your name on it. In practice it's more like a set of copies-some under your control, some under someone else's, many created automatically as a side effect of normal life. The minute your transactions leave your bank and move through an intermediary, the number of places they can live multiplies. Even if everyone behaves honorably, more storage locations mean more risk: more systems to secure, more logs, more backups, more employees with access, more vendor dependencies, more opportunities for a breach that isn't "your bank got hacked" but "a vendor three steps away got hacked."

Security is only half the story. The other half is *scope*. Not all financial data is equally sensitive, and not all apps need the same access. A simple budgeting dashboard might genuinely only need read-only access to balances and transactions. A bill-pay tool might need the ability to initiate payments. A lending product might want income verification. A "financial health" app might ask for everything because it's easier to build a product when you can hoover up data and decide later what to do with it.

The consumer experience often collapses these distinctions into a single moment: *Allow?* Yes/no. If you hit yes, you may be granting access that is broader than the mental model you had in your head. The mental model is usually: "This app will look at my transactions." The reality can be: "This app (and its vendors) can pull new transactions repeatedly, potentially across multiple accounts, for an indefinite period, and store derived insights."

Derived insights are the part people underestimate. Even if an app never sells your raw transaction list, it can still generate profiles: average balance, income volatility, likely rent amount, likely employer, propensity to overdraft, debt payments detected, subscription density, and financial stress signals. Those profiles are valuable because they're machine-friendly. They can be used to target products ("You qualify for an advance"), to tune nudges ("Increase savings now"), or to rank you in risk models. Your raw data is your autobiography; derived data is the version that fits neatly into a spreadsheet.

There's also a strange asymmetry in how consent feels versus how consent works. Consent feels like a moment. Consent works like a *standing instruction*. You don't re-approve every time the app refreshes. You approved once, and the pipeline keeps flowing. That's efficient. It's also why people can't answer a simple question a year later: "Which apps have access to your bank data right now?" If you can't list them, you don't actually have a clear perimeter.

Perimeters matter because financial data doesn't stay financial. A transaction stream can be combined with device identifiers, location data, browsing behavior, and ad-tech profiles to create a composite picture that is more identifying than any single piece. This is one reason regulators and courts have been increasingly interested in the way data brokers and adtech ecosystems handle sensitive data. A list of purchases plus a persistent identifier can turn "anonymous" into "obvious" faster than most people expect.

Even if no one is trying to identify you, identification tends to happen anyway because it's useful. Systems want to prevent fraud, link accounts, personalize offers, and reduce customer support costs. The fastest way to do that is to connect the dots. Once the dots are connected, privacy is no longer a binary state; it's a gradient of how many dots exist and how many entities can see them.

This is the honest bargain: convenience comes from letting software see more, more often. The question isn't whether that's good or bad. The question is whether you are *buying* that convenience at a price you intended to pay.

A practical way to approach this is to treat bank linking as a financial decision, not a tech decision. When you link accounts, you're increasing your exposure in exchange for value. That value might be high! A tool that prevents overdrafts, helps you budget, or makes it easier to switch financial providers can be worth it. But "worth it" should be a conscious purchase, not a default.

To make this concrete, it helps to separate what you *think* you're sharing from what might actually be shared or inferred.

Notice the word *depending*. Permissions matter. Implementation matters. Policies matter. And because most people don't read permission screens the way they read a lease, "depending" often becomes "surprise."

There's a second, quieter trade-off: accuracy versus control. When an app has direct data access, it can keep your dash-

What people think "Connect your bank" means	*What can realistically be accessed or inferred (depending on permissions)*
"My app can see my current balance."	Balances across linked accounts; average daily balance; low-balance frequency; cash buffer size inferred from patterns.
"My app can see my purchases."	Full transaction history for a period; merchant names and locations (sometimes); recurring payments; subscription stacks; bill timing; cash-flow volatility.
"It's just for budgeting."	Income detection and employer inference; debt payments and loan servicing patterns; financial stress signals; likelihood of overdraft; propensity models for marketing offers.
"I can always unlink later."	Unlinking may stop new pulls, but stored data and derived insights may remain under retention policies; multiple parties (app + aggregator) may have separate disconnect steps.

board current. When it doesn't, you end up manually entering transactions, which is tedious and error-prone. That tedium is exactly why data access exists. But the more you outsource the bookkeeping, the easier it is to forget that the bookkeeping is happening *somewhere*.

This connects to a subtle behavioral risk: once a dashboard is automatic, it becomes authoritative. People start trusting the categories, the forecasts, the "safe-to-spend" number-because it looks objective. Yet the dashboard is only as good as the data feed and the categorization model. If your aggregator mislabels transfers, your income might look smaller than it is. If reimbursements are read as "income," your savings algorithm might get too confident. If pending transactions are treated as final, your system might overreact. Automation doesn't eliminate errors; it makes them quieter.

Some of the best consumer protection in this space isn't a new law or a new encryption method. It's a simple habit: periodically checking what has access to your accounts and pruning what you no longer use. The reason this works is boring and profound: the fewer pipes you have, the fewer leaks you can suffer.

If you want a more nuanced stance than "never link anything," focus on *data minimization*. Data minimization is the idea that you should share the smallest amount of data needed to get the value you want. It's a privacy principle, but it's also just good risk management. A budgeting app doesn't need the ability to move money. A net worth tracker may not need your full transaction history; it might only need balances. A lending pre-qualification tool doesn't need permanent access; it needs a one-time verification.

Sometimes the product won't give you fine-grained choices. That's a signal too. If the only option is "all or nothing," the company is telling you something about their priorities: they've optimized for frictionless growth, not for precise consent.

Then there's the uncomfortable question: what about "free" apps? Free often means subsidized, and subsidy often means monetization. Monetization can be perfectly legitimate-referrals, partnerships, premium tiers, interest spreads. But whenever money is made from engagement, there's an incentive to keep the pipes connected, keep the dashboard sticky, keep the alerts frequent, keep the nudges persuasive. The business model doesn't automatically make the tool bad. It does mean you should expect the tool to behave like a business, not a therapist.

You don't need to become a privacy monk to be smart here. You need to become a *permission realist*. The same way you'd be cautious about giving a stranger a copy of your house key, you can be cautious about giving an app a standing invitation to read the most revealing ledger you've ever created.

One last human point. privacy isn't just about hiding. It's about dignity and options. If your data profile gets rich enough, it can shape what offers you see, what rates you get, what frictions you face, and which mistakes become expensive. The more automated finance becomes, the more your data shadow can function like a second financial self—one that follows you around quietly, influencing outcomes while you're busy living your life.

Actionable takeaway: Open your bank's "connected apps" or "third-party access" page (many banks list linked services, tokens, or data-sharing connections), and make a written inventory of every app or service with access. For each one, decide whether you still get real value *this month*. Revoke access for anything you don't actively use, and when you do link something new, choose the narrowest permissions available (read-only when possible, shortest access duration when offered). Treat bank linking like handing out spare keys: fewer copies, clearer purpose, regular re-checks.

1.4 Platform Incentives: Why Is This Free?

A free financial app is like a free drink in a casino: it's not a gift, it's a design choice.

You can feel it in the little moments. The app celebrates when you "unlock" a new feature. It puts confetti on your screen for opening an account. It nudges you to "finish setting up" direct deposit like you're abandoning a half-built IKEA dresser. None of that is accidental. A company doesn't spend money to acquire you, onboard you, and keep you engaged because it enjoys philanthropy. It does it because your attention, your deposits, your transactions, your investment orders, your insurance quote requests—some combination of those—turn into revenue.

The tricky part is that the revenue often arrives through channels that don't feel like spending. You don't swipe your card and see "App Fee: $9.99." You see "recommended," "featured," "boost your savings," "get cash now," "pre-approved," "limited time," "you qualify." The app looks like a helpful friend. Under the hood it behaves like a marketplace, a broker, or a toll road.

That's not inherently evil. In fact, some of the best consumer products in finance are "free" because they're subsidized in ways that really can lower costs. But you don't want to be naïve about the subsidy. Subsidies come with strings, and in fintech those strings are often tied to *what you do next*. The moment you understand what a platform gets paid for, the recommendations start to look less like destiny and more like sales.

Think of a platform as a place where money and data pass through, and where every pass is an opportunity to skim a little value. Not in the cartoon sense of stealing, but in the business sense of taking a spread, earning a fee, collecting a commission, or capturing a behavioral loop. Platforms are built to win by being in the middle. If you've ever wondered why an app keeps expanding into new products—banking plus investing plus credit plus insurance plus taxes—it's because the middle is where the economics get delicious.

The central question isn't "Is the app lying?" It's "What does the app get paid *more* for?" A platform can recommend ten reasonable options and still be biased toward the one that pays it best. The bias can be subtle: the paying partner shows up first, has a friendlier design, or is framed as "popular." In finance, "popular" often means "profitable."

Start with the most widespread hidden engine: card economics. When you use a debit or credit card, the merchant typically pays fees to accept that card. Part of that fee is interchange—money that flows through the payments ecosystem. Consumers experience the card as free, but merchants experience it as a cost of doing business, and those costs can shape prices and product design. This is why "cash discount" signs exist, why some stores push debit, and why certain fintech products love to become your default spending card. Transactions are not just events; they're revenue.

Once you see that, the "free" budgeting app that nudges you to use *its* card starts to make sense. It's not just offering convenience. It's trying to move your daily spending onto rails that generate income for the company. A card isn't merely a payment tool; it's a tiny toll booth that collects a fraction of your life.

The incentives become even sharper when you look at lending products packaged as lifestyle upgrades. Overdraft coverage. "Instant cash" advances. Early paycheck access. Buy-now-pay-later at checkout. These can solve real problems—timing mismatches, emergency expenses, cash flow stress. But the platform's profit often increases when you repeatedly need the product. The best customer from a revenue perspective is not the one who borrows once in a crisis and never again. It's the one who makes borrowing a habit.

That's why BNPL (buy now, pay later) is such a revealing case study. At the point of purchase, the product feels like a magic trick: split a cost into smaller pieces, often with low or no interest, and you get the item today. The monetization can come from merchant fees (the merchant pays for higher conversion), from late fees, from interchange on a linked card, or from nudges that encourage repeat usage. The consumer sees "flexibility." The platform sees "conversion," "retention," and "lifetime value." Same transaction, different emotional framing.

Notice how the interface trains your brain. A $200 purchase becomes "four payments of $50." The pain is smaller, so you buy more. Your digital balance sheet—what your accounts will look like over the next month—gets more complex. Your cash flow becomes more brittle. Then you're offered a tool to manage that brittleness, often by borrowing again. The loop can be gentle and still be a loop.

Now zoom out to "recommendations." A lot of fintech platforms monetize by being paid when they send you somewhere else. That's referral revenue, affiliate marketing, lead generation—whatever name makes it sound like a handshake instead of a commission. You click a "high-yield savings" suggestion and open an account; the platform gets paid. You request an insurance quote; the platform gets paid. You refinance a loan; the platform gets paid. The consumer experience is "comparison shopping." The business experience is "customer acquisition."

Again, this can be good. Referral-driven comparison can increase competition and lower prices. But it also creates a predictable distortion: products that don't pay referral fees may be underrepresented, even if they're better for you. Or the app may steer you toward products with higher payouts, wrapped in language that sounds objective. The most profitable product doesn't have to be a scam. It just has to be *slightly* worse for you and *slightly* better for the platform—a tiny wedge, repeated across millions of people.

The best defense is not cynicism. It's literacy. You don't need to assume every recommendation is corrupt. You just need to assume every recommendation has a business model attached, and you should be curious about it the same way you're curious about ingredients in food. If something is "sponsored," that's not automatically disqualifying—it's simply a label that tells you who is paying for the placement.

Some incentives are even more invisible because they come from *interest* rather than fees. Many fintech "cash management" products make money on the spread between what they earn on balances and what they pay you. If an app is holding your cash, your cash is not idle to them. It is inventory. That inventory can be invested (within rules), swept, or otherwise used to generate yield. If the app pays you 4% and can earn 5%, the difference is revenue at scale. That's why platforms love deposits. Deposits are sticky, predictable, and scalable.

This is also why "free" can sometimes be a trap in reverse. A platform may offer a great rate to bring money in, then quietly reduce it later when you're habituated. Or it may make the best rate conditional on behaviors that increase revenue: using the card, setting up direct deposit, making a certain number of transactions, subscribing to a premium tier. The product is not just a bank account; it's a behavior funnel.

Investing apps provide a particularly clear example of how incentives shape design. Commission-free trading feels like consumer liberation—no more $9.99 trade fees. But "free trading" has to be funded somehow. One mechanism is payment for order flow (PFOF), where a brokerage may receive compensation for routing customer orders to particular trading firms. The details can get technical quickly, but the consumer-relevant point is simple: when you're not paying explicitly, you should ask where the money comes from and whether it could affect execution quality.

This is where conflicts of interest live. A broker has a duty to seek best execution—meaning it should try to get you a good outcome on price and speed, not just route your trade wherever it's most profitable for the broker. Regulators and industry rules focus heavily on disclosure and supervision for exactly this reason: the incentive to monetize order routing exists even when the app looks like a neutral tool. You don't have to become a market-structure nerd to be protected. You just have to recognize that "free" can be funded by mechanisms you never see on a receipt.

The more general principle is that a platform will often optimize for what it can measure and monetize. Engagement is measurable. Transactions are measurable. Referrals are measurable. Holding balances is measurable. Long-term well-being is harder to measure, and it's rarely paid for directly. This is why some apps are obsessively chatty about daily streaks and tiny wins while being oddly quiet about big risks like concentration, insurance gaps, or debt spirals. Big risks are bad for the brand and sometimes bad for conversion.

A "nudge" is where these incentives become personal. Nudges are the little design choices that steer you without forcing you: the default option, the color on the button, the placement of the "skip" link, the timing of the notification. Nudges can help you do what you already want (save more, pay bills on time). They can also help a platform do what it wants (get you to open another product, keep you moving money through its pipes, increase the number of monetizable events).

The tension becomes sharper because many nudges are genuinely helpful. That's what makes them powerful. If an app saves you from a late fee once, you start trusting it. Trust is the real currency. After that, the app can ask you to take bigger steps: "Try our credit builder," "Switch your direct deposit," "Get cash instantly," "Invest your spare change," "Upgrade for premium insights." Each step might be fine. The question is whether the sequence is building your financial resilience—or building the platform's revenue stability.

You can often spot incentives by noticing what the app celebrates. Some apps celebrate "net worth up." Great. Others celebrate "you made 23 transactions this month." That's not a financial milestone; it's a business milestone. Some apps celebrate "you got approved." Approval is not a virtue. It's a pricing decision.

Another tell is when recommendations are oddly specific. "Based on your profile, we recommend this card." Maybe. But also: the platform can only recommend products it has a deal with, products it can technically integrate, or products that fit its risk tolerance. The "market" you see inside an app is not the full market. It's a curated shelf.

Curated shelves aren't bad. Grocery stores curate shelves too. The difference is that groceries are tangible and the trade-offs are obvious. Finance products are abstract, and the trade-offs are often delayed. A bad credit product doesn't taste bad in the moment. It tastes bad three months later when interest accrues, when late fees stack, when your cash flow tightens. Platforms know this. That's why the marketing is so heavily front-loaded in "ease" and "instant."

So how do you protect yourself without turning every financial decision into a courtroom drama?

Start by treating "free" as a prompt to ask one calm question: *What behavior is this product trying to increase?* Is it trying to increase deposits? Spending? Borrowing? Trading? Referrals? Data connections? The answer doesn't automatically disqualify the product, but it tells you where the platform's loyalties will drift when your interest and their interest diverge.

Then translate recommendations into numbers you can compare. If it's a savings product, compare the APY and the conditions. If it's a credit product, compare the APR, fees, and what happens if you miss a payment. If it's a trading product, look at order type options, price improvement disclosures if available, and how the broker describes its routing and execution practices. If it's insurance, compare coverage terms, not just monthly premium. Incentives love vagueness; consumer protection loves specifics.

Finally, remember the lesson from the plumbing: timing and friction are where people get hurt. Platforms often monetize by reducing friction at the front end—one-click approvals, instant access, seamless onboarding—and letting the complexity show up later in fees, fine print, or behavioral loops. Your job is to reintroduce a little healthy friction for yourself: a pause, a comparison, a second quote, a quick search outside the app's walled garden.

Actionable takeaway: The next time a financial app recommends a product, don't accept or reject it immediately. Instead, write down two lines in your notes app: (1) *What does the platform get paid for if I do this?* (referral fee, interchange, interest spread, order routing, subscription upgrade, data access) and (2) *What is the downside if I'm wrong?* (fees, rate changes, lock-in, hard credit pull, cash-flow stress). If you can't answer line (1) from the app's disclosures, treat that as a reason to slow down and look it up elsewhere before clicking "Apply."

Chapter 2

How Machines Think About Money

Imagine teaching a dog to sit. Does it understand the command, or did it just notice you twitch your hand? Financial AI is similar: it finds patterns, not meaning. Before we trust these systems with our retirement or insurance, we need to understand that they aren't magic-they are math. This chapter pops the hood on the algorithms driving your financial life, revealing why 'smart' machines make strange mistakes and why a computer that can beat you at Chess might still be terrible at managing your portfolio.

2.1 AI vs. Automation: Who Is Driving the Car?

The most expensive misunderstanding in modern finance is calling everything *AI*. Your paycheck hits, your rent gets paid, your credit card autopays, your robo-adviser rebalances, your bank flags a charge in another state, your insurer reprices your premium, and somewhere in the blur your brain files it all under one label: *the computer decided*. But there are two very different creatures hiding inside that phrase. One is a rule-following clerk who never improvises. The other is a pattern-hunter who *must* improvise, because it's trying to predict what you'll do next.

If you can tell which one you're dealing with, you can predict the kinds of mistakes it makes, the kinds of questions you should ask, and the kinds of damage it can do to your wallet. If you can't tell, you'll argue with the wrong thing. You'll demand an explanation from a system that can only shrug, or you'll trust a system that *sounds* smart but is really just doing what somebody coded years ago and forgot to revisit.

Start with a familiar moment: it's Thursday night, you open your banking app, and you see a payment went out to your student loan servicer. You didn't tap anything. You didn't approve anything. You didn't even remember it was due. Still, the payment happened. That's not a crystal ball. That's not a model. That's not "learning." That's automation: a precommitted instruction carried out on schedule. The banking app is basically a wristwatch with a button.

Automation is the oldest kind of financial "intelligence": *If this, then that*. If the date is the 15th, pay the bill. If the balance drops below $300, move $200 from savings. If the price falls 10%, sell. If the fraud system sees three transactions in five minutes, lock the card. You can make automation complex, but it's still fundamentally a recipe. The key point is that the recipe already contains the decision. The system isn't guessing. It's executing.

AI, as most financial firms use the word today, is different in a way that matters emotionally. It doesn't feel like a recipe. It feels like a judgment. Machine learning systems ingest piles of data and spit out a prediction: *This borrower will probably default. This driver will probably file a claim. This customer will probably close their account. This market regime looks like the ones that preceded drawdowns*. And then a business process turns those predictions into actions: deny, approve, price higher, rebalance, route to a human, send a warning, quietly reduce limits, quietly increase scrutiny.

If automation is a light switch, machine learning is a motion sensor. A light switch can be annoying in a predictable way: it's either on or off, and if it's wrong, you fix the setting. A motion sensor can be annoying in a personal way: it decides you are "not really there" when you're sitting still reading, and it decides you "are there" when your cat wanders by at 2 a.m. Neither is evil. But only one is making guesses about the world.

That "guessing" word is worth lingering on, because it's the source of both the magic and the misery. Machine learning doesn't usually output a single crisp decision. It outputs a probability, a score, a ranking, a confidence level. Somewhere behind the scenes, a threshold turns that score into a yes/no action. That threshold might be different at 9 a.m. than at 9 p.m. It might tighten when markets get jumpy or when a lender's loss rates tick up. It might loosen in a growth push when a company wants more customers. A person can change the threshold without changing the model and claim the "AI" did it. A person can keep the threshold and change the model and claim the "policy" did it. You, on the other side of the app, just feel the door open or slam shut.

A good way to keep your footing is to ask one plain question: *Is this system following instructions, or is it inferring intent?* Automation follows instructions. AI infers intent. Automation is comfortable with certainty: it only needs to know what time it is or whether a condition is met. AI lives in uncertainty: it tries to estimate what you are likely to do, based on patterns in data that may include you, resemble you, or be "about you" without being *from* you.

Here's a simple example that exposes the difference. Suppose you set your credit card to autopay the minimum due. That's automation. The bank isn't judging your character; it's obeying your preference. But suppose the bank decides to lower your credit limit because a model predicts you are more likely to miss payments in the next six months. That's AI (or at least machine learning) wrapped in policy. Both actions change your financial life. One is a contract you initiated. The other is a forecast the institution initiated.

The confusion gets worse because companies love to market automation with AI language and AI with automation language. "Smart transfers." "Intelligent budgeting." "AI-powered approvals." Sometimes the product is genuinely adaptive. Sometimes it's a rule engine with a new coat of paint. The marketing copy doesn't help you, because the goal isn't conceptual accuracy; it's emotional comfort. "AI" sounds modern. "Automation" sounds dull. But dull is sometimes exactly what you should want when money is involved.

The real difference shows up when things go wrong. Automation fails like a toaster: it burns the bread the same way every morning until someone changes the setting. AI fails like a weather forecast: it can be impressively accurate for long stretches, then suddenly wrong for reasons that are hard to explain in the moment, because the world changed in a way the system didn't anticipate. If you expect a toaster and you get a forecast, you'll feel betrayed. If you expect a forecast and you get a toaster, you'll feel gaslit: "Why can't it understand what I mean?"

Take the classic personal-finance automation trap: the overdraft cascade. You set your utilities to autopay, you set your credit card to autopay, and you set your streaming subscriptions to autopay. Each instruction is reasonable in isolation. But automation doesn't understand your life; it understands its checklist. If a paycheck lands a day late, a half-dozen payments can hit before the deposit, and suddenly you're paying overdraft fees that feel like punishment for being organized. The system didn't predict the paycheck delay, didn't weigh which bills were most important, didn't attempt to minimize harm. It simply executed.

Now imagine the AI-flavored version of the same pain. Your bank has a model that tries to predict when you're likely to overdraft. If it thinks you're headed toward trouble, it might push you a warning, offer a small cushion, or move money between accounts if you've opted in. Helpful. But it might also do less visible things: tighten transaction approvals, decline a borderline debit, hold a deposit longer under a risk policy, or route your customer support request into a slower queue. The point isn't that these actions are necessarily malicious. The point is that AI can create invisible friction based on a guess, and invisible friction feels like the world is suddenly biased against you.

The "who is driving the car?" question becomes urgent because finance is not a sandbox. A small change can have second-order effects. A lowered credit limit can raise your credit utilization ratio, which can ding your score, which can increase your loan APR, which can change your insurance premium in some states and contexts, which can reduce your monthly slack, which can increase the likelihood of missing a payment, which can confirm the model's original suspicion. Automation can create chains too, but AI is especially good at creating feedback loops because it continually watches outcomes and updates what it believes about you.

To make this concrete, picture two different drivers sitting behind the wheel of your finances.

The automation driver is literal-minded. If you told it "keep $2,000 in checking," it will try to keep $2,000 in checking even if you have a $1,900 car repair and a $2,100 rent payment back-to-back. It won't ask if you *meant* the rule to flex in emergencies. It can't feel embarrassment when it transfers money at the worst possible moment. It's not trying to be fair or helpful. It's trying to be consistent.

The AI driver is improvisational. If you told it "keep me safe from overdrafts," it might start behaving like an anxious friend: warning you early, nudging you often, and sometimes being wrong in ways that feel personal. It might see patterns you didn't realize you were broadcasting (how quickly you spend after payday, how often your balance dips below a certain line, how your transactions change around holidays) and then act on those patterns. And because it's predicting, it can be wrong *confidently*.

One reason financial AI feels spooky is that the input data doesn't look like "money." People imagine a bank model reading their income and bills like a neat spreadsheet. In reality, modern systems can incorporate thousands of features: timing, frequency, category mixes, device fingerprints, location consistency, merchant patterns, typing cadence, transaction velocity, even which menus you tap in an app and how long you hover before canceling. Some of this is used for security, some for marketing, some for risk, some for "experience." From the company's perspective, it's just signal. From your perspective, it's the digital equivalent of a stranger saying, "I can tell you've been stressed lately," because you bought antacids and slept irregularly.

This is where the language matters. Automation is generally legible: you can point to the rule. AI is often statistical: you can point to the outcome and maybe a few influential factors, but you can't point to a single human-readable instruction that says, "Deny this person because they bought more gas than usual in the last three weeks." That isn't how the system "thinks." It doesn't reason in sentences. It associates patterns with outcomes. It's closer to a very elaborate hunch machine.

Even within the same company, you may be interacting with both at once. A robo-adviser might use automation to rebalance your portfolio on a schedule, but use a machine learning model to decide which clients get proactive calls during volatile markets. A lender might use a model to generate a credit risk score, then use automation to apply policy rules: if score is below X, deny; if score is above Y, approve; if in the middle, request more documentation. A fraud system might use a model to rate a transaction's likelihood of being fraudulent, then an automated policy decides whether to decline, text you, or allow.

That blending is why "AI versus automation" isn't really an argument about labels. It's a question about *where discretion lives*. In a fully automated rule system, discretion happens at design time: someone writes the rules. In a machine learning system, discretion happens at run time: the model applies learned patterns to new situations. The model is, in effect, making a fresh judgment every time it sees new data.

That discretion can be valuable. Consider people with "thin files" in credit: not much traditional history to score. Alternative signals (used carefully) can help lenders identify reliable borrowers who would otherwise be invisible. Or think about fraud: the best defense against thieves is not a list of last year's scams; it's a system that can adapt when thieves adapt. In these use cases, "guessing" is a feature. A rigid checklist is easy to game and easy to outgrow.

But discretion is also how you get the maddening experience of being treated like a different person overnight. The same spender, the same bills, the same income—yet a bank suddenly wants you to verify your identity, an insurer suddenly raises your premium, a payments app suddenly freezes a transfer, an investment platform suddenly asks you to re-answer risk questions. You didn't get worse. Their model got more nervous. Or their training data shifted. Or the threshold changed. Or the company decided it wanted different customers this quarter than last quarter.

If you want a mental model that's both accurate and oddly calming, try this: automation is *obedience*, AI is *probabilistic caution*. Obedience can be brittle. Probabilistic caution can be unfair. Neither is inherently virtuous, and both can be weaponized by incentives. Companies don't deploy these systems because they love you. They deploy them because prediction and scale are profitable. Sometimes those incentives line up with your interests (catching fraud, lowering costs, making basic services accessible). Sometimes they don't (steering you toward higher-margin products, pricing you more aggressively, quietly reducing service).

This is why the "driving" metaphor matters. When you're in the passenger seat, you don't just care whether the car stays on the road. You care about what the driver does in edge cases: heavy rain, unexpected detours, a child running into the street. That's where philosophy turns into outcomes. A rules-based system is predictable in edge cases, but predictably dumb. A model-based system is flexible in edge cases, but flexibly wrong.

Here's a quick test you can run in your own financial life to identify which one you're dealing with: *Can you change the outcome by changing the instruction?* If you can fix the problem by editing a setting ("pay on the 28th instead of the 25th," "move $50 instead of $200," "only transfer if balance is below $100"), you're mostly dealing with automation. If the outcome seems to change even though you didn't change a setting—especially if it changes in response to your behavior over time—you're likely dealing with an AI/ML model somewhere in the chain.

Another test: *Does the company talk about "learning," "signals," or "patterns"?* That's a tell. "Learning" usually means the system is trained on historical examples. "Signals" means inputs may be indirect. "Patterns" means the decision is coming from correlation rather than a straightforward rule. None of this is automatically bad; it's simply a warning label that you're not negotiating with a checklist.

A third test, more emotional than technical: *Does the decision feel like it's about you?* Automation errors feel mechanical. AI errors feel like a judgment call. When your card declines because the terminal is offline, you blame the plumbing. When your card declines because "unusual activity" is detected and you were just buying groceries like always, you start imagining an invisible dossier. That sense of being profiled is not paranoia; it's an accurate description of what predictive systems do. The question is whether the profiling is bounded, accurate, contestable, and governed—or whether it's a fog you can't see through.

People often ask whether AI is replacing humans in finance. That's the wrong fear. The more common reality is that AI is *replacing the human explanation*. Humans are still involved—designing, selecting data, setting thresholds, choosing objectives, approving deployment, responding when things break. But many frontline interactions now happen through a score and a script. When you call to ask "why," the human you reach may not know. They may only see the decision and a few canned reason codes. And if the system is a complex model, the organization may have traded away the ability to give you a satisfying narrative in exchange for a small improvement in prediction accuracy.

This trade—accuracy for explainability—shows up everywhere. In investing, a simple rebalancing rule is easy to describe and audit: "If stocks exceed 60%, sell down to 60% and buy bonds." A machine learning strategy might say it's detecting "market regimes" and "optimizing factor exposures." That can be real. It can also be a fancy way of saying, "We found a pattern in the past and we're betting it continues." Either way, your ability to evaluate the strategy changes. With a rule, you can ask, "Do I agree with this rule?" With a model, you have to ask, "Do I trust the process that produced this model, and do I understand how it fails?"

Finance has always run on models, even before anyone said "AI." A mortgage calculator is a model. A credit score is a model. A retirement projection is a model. What machine learning changed is the *texture* of the model: more inputs, less obvious logic, more adaptation, more reliance on patterns that are not common sense. The old model world felt like math. The new model world feels like behavior.

And behavior is messy. People change jobs. People get divorced. People get sick. People move. People panic-buy at the top and swear they'll never invest again at the bottom. Markets change too. A model trained on a calm decade can behave like a nervous animal in a storm. Automation doesn't have this particular problem; it doesn't "learn" a calm decade. It just executes. That's why automation can be a safety feature in your own financial habits—steady contributions, steady bill payments—while being a risk in environments where the world changes faster than your rules.

So what does a financially literate person do with this distinction, besides nod and move on?

You treat automated systems like contracts and machine-learning systems like opinions. Contracts require careful setup and periodic review. Opinions require skepticism and the right to ask "based on what?" If your bank or app can't tell you whether a feature is rule-based or model-based, that's not a minor transparency issue; it's a practical obstacle to self-defense.

A useful stance is to demand *two kinds* of clarity. First: *What did I authorize?* If it's automation, you should be able to see the exact triggers and limits. Second: *What did the system infer?* If it's AI, you should be able to see the major categories of data used and the type of prediction being made (fraud risk, default risk, churn risk, identity risk). You don't need a PhD-level explanation; you need a consumer-level map of what's driving the car.

The actionable takeaway is simple and immediately usable: the next time a financial app offers a "smart" feature, force it into one of two buckets before you turn it on. Ask, out loud if you have to: *Is this a rule I control, or a prediction they control?* If it's a rule, set guardrails (caps, dates, minimum balances) and schedule a quarterly check-in on your calendar to make sure the rule still matches your life. If it's a prediction, assume it will sometimes be wrong, keep an escape hatch (a human support path, a backup payment method, a second bank account), and avoid giving it permission to move money or change commitments without your explicit confirmation. That's how you stay the driver, even when the car is full of computers.

2.2 The Rearview Mirror Trap: Why Models Fail

A backtest is the financial equivalent of a highlight reel. It has the same seduction and the same flaw: it edits out the part where you didn't know what was going to happen next. When an investing app boasts that its "AI strategy" would have turned $10,000 into $67,000 over the last decade, it's showing you a perfectly lit stage with the wind at the performer's back. Real investing is performed in the dark, with the floor moving.

The uncomfortable truth is that financial AI is mostly a specialist in yesterday. It's trained on historical data, validated on historical data, tuned on historical data, and marketed with historical data. Even the language gives it away: "historical accuracy," "track record," "proven in past markets." That's not necessarily a scam; it's the nature of the beast. You can't train a model on the future. You can only train it on a fossil record of human behavior and market structure that may never reappear in quite the same shape.

That's the rearview mirror trap: the model drives by looking backward. Not because it's stupid, but because it has no other view. It can measure patterns in the data it has, then assume those patterns persist long enough to be worth money. In markets, that assumption is always on probation.

The easiest way to see the trap is to imagine teaching someone to drive using only recordings of past traffic. They'd learn that red lights are usually followed by green lights, that cars generally stay in lanes, that rain makes people slow down. They'd ace the exam if the exam is made from the recordings. Then you put them on a real highway during a snow squall, with a lane closure and an ambulance weaving through, and suddenly their "accuracy" becomes a liability. They don't know what matters when the world stops behaving like the training set.

Machine learning has a name for the most common version of this failure: *overfitting*. Overfitting is what happens when a model doesn't learn the underlying skill, but memorizes the practice test. It's not always obvious, because memorization looks like competence when you're grading the same material it memorized.

In investing, overfitting is often dressed up as sophistication. A simple strategy might use a handful of sensible signals: valuation, momentum (recent performance), maybe interest rates. An overfit strategy uses everything it can get its hands on: dozens of indicators, interactions between indicators, timing rules, filters, "regime detectors," and tweaks that seem small but are actually fingerprints of the past. It can become less like a compass and more like a lock that only opens for one specific key: the specific sequence of market events that already happened.

There's a particular kind of heartbreak that comes from overfitting because it feels so *earned*. Someone worked hard on that model. They tried hundreds of variations. They fought to reduce errors. They polished the backtest until it fit like a tailored suit. And then, the moment the model meets new data, it behaves like a person who memorized the answers but never learned the subject. When the question is asked a different way, it freezes-or worse, it answers confidently and incorrectly.

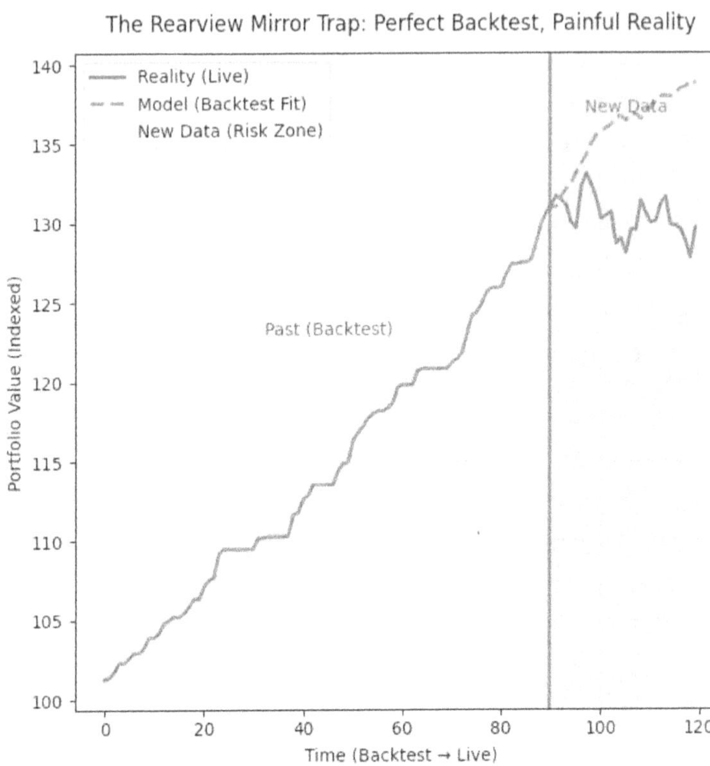

The chart is cartoonish on purpose. In the real world, models don't usually fail with that kind of dramatic honesty. They fail in a more dangerous way: *they slowly stop working while still sounding sure of themselves*. The backtest keeps glowing in the app store screenshots. The model keeps producing numbers. The portfolio keeps trading. The only early warning sign is that the story you were sold—"it adapts," "it learns," "it sees patterns humans miss"—starts feeling like a slogan rather than a description.

The reason this happens isn't mystical. Markets are not physics labs where the same experiment can be repeated under the same conditions. Markets are crowds with money. Crowds notice strategies. Crowds copy strategies. Strategies change behavior, and behavior changes the market the strategy was trained to understand. In other words, financial models can destabilize the very patterns they're exploiting. The act of predicting can change the thing being predicted.

Even if nobody copies your model, the world still changes. Interest rates move. Regulations shift. A new trading venue becomes dominant. A pandemic scrambles supply chains. A war changes energy prices. A central bank surprises everyone. Sometimes the change is slow, like a climate shift. Sometimes it's a cliff. In machine learning terms, the data-generating process changes. In plain English, the game starts being played by new rules.

That kind of change is often called a *regime change*. You don't need to use the term, but you've felt it. A "normal" day in markets suddenly has 3% moves. The stocks that used to cushion your portfolio start falling in sync with everything else. The safe asset isn't safe for months at a time. Correlations-the way different investments move in relation to each other-rearrange themselves like furniture in the dark.

Models trained on calm periods tend to be optimists. They learn that dips are usually followed by recoveries, that volatility is usually mean-reverting, that liquidity is usually there when you need it. When the environment changes, those learned comforts become wrong precisely when the cost of being wrong is highest.

There's a deeper reason the rearview mirror trap is so sticky: it's incredibly easy to fool yourself with historical data. Not always intentionally, either. If you try enough variations of a strategy, *some* variation will look brilliant purely by chance. That's not insight; it's statistical luck dressed up as skill. The more knobs you turn, the more likely you are to find a configuration that perfectly explains the past-even if it has no causal connection to the future.

This is why "more complex" doesn't automatically mean "more accurate" in finance. Complexity can be a liability when the historical record is short or the environment is unstable. A model with a thousand parameters has a thousand ways to be accidentally right yesterday and wrong tomorrow. Simpler models can be harder to impress people with, but they are often harder to break. They're like a hammer: not elegant, but reliable across a lot of messy situations.

Overfitting also shows up in the way products are packaged. A strategy isn't just a model; it's a *set of business decisions around a model*. Which data was included? Which costs were assumed? Which trades were assumed possible? How much money was assumed to be deployable before returns deteriorate? Did the backtest include realistic transaction costs and taxes? Did it assume you always get filled at the desired price? Did it quietly ignore the fact that the strategy would have been impossible for an individual investor to execute at the time?

A backtest can be "accurate" and still be irrelevant if it assumes frictionless trading in a world that charges commissions, spreads, and taxes. That's not a minor accounting detail; it's the difference between a strategy that works on paper and a strategy that survives contact with reality.

And then there's the big one: incentives. If a firm is paid based on assets under management, it benefits from showing a smooth, market-beating backtest that attracts deposits. It doesn't get paid more for showing you a backtest that says, "This is fragile and will probably stop working when volatility spikes." Humans rarely market fragility. Models rarely announce it. The rearview mirror trap is profitable.

None of this means models are useless. It means you should treat them like you would treat a weather model: valuable, imperfect, and dangerous if you outsource your judgment. A forecast can help you decide whether to bring a jacket. It cannot guarantee sunshine at 3 p.m. If someone sells you a forecast as a guarantee, they're selling you a feeling, not a tool.

The difference between responsible modeling and reckless modeling often comes down to whether the designers planned for failure. In well-run financial institutions, there's a whole discipline around this idea: *model risk management*. The intuition is simple even if the implementation is not. A model can hurt you in two ways: it can be wrong, and it can be used wrong. A correct model can be misapplied outside its intended scope. A decent model can be deployed with sloppy controls. A model can be right on average and still disastrous in the tails-the rare moments you most care about.

If that sounds abstract, translate it into everyday money decisions. A spending model might correctly identify that most customers who behave like you are fine-until you lose your job, at which point the same behavior has a different meaning. A credit model might correctly score risk in normal times-until inflation forces households to juggle bills differently. An investing model might correctly ride momentum trends-until a sudden policy shock reverses everything and the model's "buy the winner" instinct becomes "buy the falling knife."

The market has its own way of punishing overconfident automation and overconfident AI: speed. When a model is plugged into a system that can trade, lend, or price in milliseconds, small errors compound before anyone can blink. This is one reason market regulators and firms care so much about testing and controls for algorithmic trading systems. A bug, a mis-specified parameter, or a model that behaves weirdly in a rare corner case can turn into real losses quickly—not because the system is evil, but because it is *fast and literal.* It doesn't hesitate. It doesn't feel fear. It doesn't stop to ask if this is a good day to take a risk.

At the consumer level, the same dynamic shows up as "instant decisions." Instant approvals. Instant rejections. Instant fraud blocks. Instant price updates. The speed feels modern, but it also narrows the window in which humans can catch errors before they affect you.

A helpful way to think about the rearview mirror trap is to separate two questions that often get mashed together: *Did the model work in the past?* and *Will it work when it matters?* The first is easy to demonstrate. The second is the whole game.

The second question depends on whether the model's success came from something durable or something accidental. Durable signals tend to have a story behind them that survives different environments. For example, it's not shocking that companies with strong balance sheets tend to be more resilient when credit tightens. That has a business logic. Accidental signals don't have that kind of story; they're patterns that show up in the data because the past has quirks. If you find a "signal" that predicts returns based on, say, the first letter of a company's ticker, you've discovered not an edge but a prank played by randomness.

Machine learning is powerful partly because it can find subtle patterns humans miss. The danger is that it can also find subtle patterns humans *should ignore*. In markets, there are infinite coincidences. A sufficiently flexible model can always find a few that look like prophecy.

This is why "99% historical accuracy" is not a flex; it's a warning. In many financial contexts, the more perfect the fit, the more you should suspect the model has latched onto noise. Real financial relationships are messy. They degrade. They drift. They are constantly being arbitraged away. A model that claims near-perfection is often telling you it has memorized the exam.

So what can you do with this, practically, without turning your life into a research project?

You can stop treating backtests as proof and start treating them as *stress stories*. When you're evaluating an investing strategy-especially one marketed as "AI-powered"-don't ask only, "What's the return?" Ask, "What's the drawdown?" (drawdown is the peak-to-trough decline). Ask, "What happened in ugly markets?" Ask, "How does it behave when bonds fall with stocks?" Ask, "What assumptions does this require to keep working?" You're not trying to win a debate. You're trying to understand how the strategy fails, because that's when you'll meet the real version of it.

And if you're using an AI-flavored product outside investing-credit, insurance, banking-the same idea holds. The impressive demo is the easy part. The meaningful question is what happens when you're the weird case: the traveler, the gig worker, the person with irregular income, the person who just moved, the person who had a medical crisis, the person whose identity got tangled up with someone else's data. Models are built for the average. Life is lived in the exceptions.

The actionable takeaway is to adopt a "rearview mirror rule" before you trust any model with real consequences: *If the pitch is a backtest, demand a failure plan.* In practice, that means you ask one blunt question-either to the firm, to the documentation, or to yourself if the answers aren't available: *What is the model expected to do when the world stops looking like the training data?* If you can't find a clear answer-if there's no mention of limits, no discussion of bad periods, no acknowledgment of regime changes, no transparency about assumptions-treat the performance claims as advertising, not evidence. Then set your own guardrails: keep diversified holdings, avoid concentrating in one "genius" strategy, and never allocate money you can't afford to watch underperform for a long, uncomfortable stretch. That's how you keep yesterday's data from driving your future off the road.

2.3 The Black Box: Why the Computer Can't Explain Itself

The email arrives with the emotional warmth of a parking ticket: *We're unable to approve your application at this time.* No voice. No face. No "here's what you could do differently." Just a decision, clean as a guillotine. If you've ever had that happen- credit card, apartment, auto loan, a "buy now, pay later" limit- you know the particular kind of frustration it creates. You're not only disappointed. You're *stuck*. You can't argue with a sentence fragment.

A lot of people assume the computer must have a crisp reason. Somewhere, they imagine, is a simple line of code: *deny if debt-to-income ratio is above 40%*. That's how the world used to work when decisions were closer to automation. But modern underwriting is often built on machine learning models that behave less like a checklist and more like a dense thicket of tiny signals. The result is the "black box" problem: the system produces a decision that can be statistically impressive, yet hard to translate into a human explanation that feels satisfying—or even coherent.

A black box isn't necessarily a sinister box. In engineering, it just means you can observe the inputs and outputs but not the internal logic in a way that's easy to summarize. You put in a pile of information about a borrower, and you get out a score: *risk of default*. You put in details about a claim, and you get out a score: *fraud likelihood*. You put in a transaction, and you get out: *approve/decline*. The system may be doing something perfectly rational in its own mathematical language, but it can't sit down across from you and say, "Here's what I noticed, and here's why it matters."

That mismatch—between machine logic and human conversation—isn't a minor UX issue. It's the point where "AI in finance" collides with your rights, your time, and your dignity. If a model can't explain itself, it's harder to correct mistakes, harder to contest unfairness, and harder to plan your next move. It turns a financial system that should feel like a set of rules into a fog bank.

To understand why this happens, it helps to be blunt about what many machine learning models actually are: extremely complicated curve-fitting machines. They ingest thousands of examples-past borrowers, past delinquencies, past fraud cases-and they learn a function that maps inputs to outputs. The model is rewarded for being correct, not for being articulate. In many cases, the model doesn't contain a neat list of "reasons." It contains weights, splits, and interactions that, collectively, produce a probability.

If you asked a model, "Why was I denied?" it might answer, honestly, "Because when I combine 600 small cues in the way I was trained to combine them, the number ends up on the wrong side of a threshold." That's not a satisfying explanation, but it's closer to the truth than the comforting fiction that there is *one* decisive factor.

This is the part that makes people angry, and not unreasonably: the model is being treated like a decision-maker, but it can't be cross-examined like one. Humans can be biased, petty, inconsistent-yet humans can be interrogated. A human can be forced to articulate a rationale, and that articulation creates accountability. A black box can feel like power without accountability, which is a pretty good definition of what consumers fear most.

Financial firms will often say they can't reveal details because the model is proprietary or because disclosing too much would help fraudsters. Those concerns are real. But they're also convenient. "We can't tell you" is a remarkably effective way to end a conversation, especially when you're the one who needs money and they're the one who has it.

Here's what complicates things: black-box models can be *genuinely* better at prediction. If the goal is to estimate default risk, a model that can pick up subtle, non-linear interactions may outperform a simple scorecard. That improved prediction can reduce losses, which can in theory lower rates for some borrowers and expand access for others. In fraud detection, black-box models can be the difference between stopping a takeover and letting your account be drained. Accuracy can be consumer protection.

But accuracy has a cost. The cost is often paid in the currency of explanation.

Think about the difference between a teacher grading an essay with a rubric versus a teacher running the essay through a complicated plagiarism detector. With the rubric, you can see what to fix: thesis, evidence, organization. With the detector, you might just get a red flag. The detector might be right. It might also be wrong. Either way, if it won't show its work, you're left with a kind of bureaucratic helplessness.

That's the black box in finance: a score that behaves like a verdict.

The most common way black boxes hurt consumers is not dramatic. It's administrative. Data errors happen constantly: a wrong address, a mixed-up identity, a late payment that wasn't late, a duplicate record, a payment plan recorded incorrectly, a disputed charge that never got tagged as disputed. When decisions are powered by models that pull from multiple data sources, errors can compound. A small mistake in one corner of your profile can ripple outward—into credit limits, insurance pricing, fraud flags, and even the level of friction you face when you try to do routine things like transfer money or open a new account.

When the decision pipeline is opaque, you can't easily locate the rotten link in the chain. You don't know which bureau, which dataset, which field, which timestamp, which "feature" mattered. So you're stuck attacking the problem with a blunt instrument: calling customer service, filing disputes, mailing documents, waiting weeks. The black box doesn't just deny you; it denies you *the map* of how to fix the denial.

There's also a subtler cost: the black box can train you to accept financial outcomes as fate. You start hearing "the system" the way people talk about weather. *Nothing I can do; it's just my score.* That resignation is a gift to any institution that would prefer you not ask questions. And it's a tragedy for consumers, because a lot of "score" outcomes are changeable—if you know what the levers are.

The good news is that finance isn't allowed to be completely mystical. In the U.S., when you're denied credit (or offered worse terms) because of information used in a credit decision, lenders generally have to provide what's called an *adverse ac-*

tion notice: specific reasons for the decision. The existence of the notice is a kind of legal refusal to accept the black box as an excuse. "Our model is complicated" is not supposed to be a get-out-of-explanation free card.

This matters because it creates a pressure valve. If you're told, "Denied due to insufficient income," you have a lever: document income, reduce obligations, apply for a smaller loan, add a co-signer (with care), or try a different product. If you're told, "Denied due to recent delinquency," you know to check your reports and dispute errors. If you're told, "Denied due to too many recent inquiries," you know to stop spraying applications like confetti. These reasons can be imperfect, but they are actionable.

The tension is that a modern model might not have a single, simple "reason" in its own internal wiring. So institutions often translate the model's output into a set of standardized reason codes: the closest human-understandable explanations that fit within regulatory expectations. That translation is better than silence, but it can still be frustrating. It can feel like you're being told, "Something about your finances made us nervous," which is not a reason so much as a mood.

And even when the reasons are accurate, they can be incomplete in the way a medical diagnosis can be accurate but not helpful. "High debt-to-income ratio" might be true, but *which* debt? "Insufficient credit history" might be true, but *what* would count as sufficient? If the institution won't clarify, you're left guessing which changes are worth your effort.

This is where the "interpretability versus accuracy" trade-off becomes personal. *Interpretability* means you can understand how a model uses inputs to produce outputs—at least enough to challenge it, audit it, and correct it. A very interpretable model might be a simple scorecard: points for steady income, points off for recent delinquencies, points off for high utilization. You can look at it and say, "I disagree with that weight," or "That factor is outdated," or "That data is wrong."

A less interpretable model might be a deep learning system with hundreds of thousands of parameters. It might produce better predictions, but it won't be able to narrate itself in plain language. You can bolt explanations onto it after the fact—tools that estimate which features were influential for a particular decision—but those explanations are approximations, not the model's own reasoning rendered into English. They're more like a detective's reconstruction than a defendant's confession.

This matters not only for you as an individual but for the institutions that deploy these systems. A model that no one can interrogate is hard to govern. Good organizations build *effective challenge* into their model process: people whose job is to poke holes, test edge cases, ask "what if," and insist on documentation that survives handoffs. If the model is too complex to explain internally, it becomes too complex to control. That's not philosophical; it's operational. If no one can explain why the model suddenly started declining 15% more applicants in a particular region, the business can't quickly tell whether it's catching fraud, accidentally discriminating, or simply drifting into error.

When you hear "AI is objective," it's worth remembering that objectivity requires traceability. You can't audit a vibe. You can audit a rule. You can audit a dataset. You can audit a model's performance and bias metrics. But the harder it is to explain a model, the easier it is for problems to hide in the seams—especially when a company is scaling fast and the incentives are screaming "ship it."

The black box also changes the emotional posture of customer service. A human loan officer can sometimes say, "Here's what to work on, come back in six months." A model-driven system might not have that kind of human discretion. The representative may genuinely want to help but only see a screen that says *Declined* and a couple of generic reasons. The institution has essentially put the employee in the awkward role of interpreter for a system that doesn't speak.

This is why contestability—the ability to challenge and correct a decision—matters so much. Without contestability, "AI" becomes a polite way of saying "final." Contestability doesn't mean you always win. It means there's a path to show the system it's wrong, or to show the institution that a policy has become unreasonable, or to fix broken data. It's the difference between a financial system that feels like law and one that feels like fate.

If you want to keep your head in a world of black boxes, it helps to separate two kinds of "why."

There's the *global why*: why does this model generally deny people? What factors does it tend to care about? What population was it trained on? How is it monitored for drift? Is it tested for disparate impacts? Is it validated independently?

Then there's the *local why*: why did it deny *me*, today, for this product, with my information? A global explanation might satisfy a regulator or an auditor. A local explanation is what you need to make your next move.

Many companies are better at global why than local why, because global explanations can be packaged. Local explanations create vulnerability: they create evidence. They create appeal rights. They create customer support burdens. They also create accountability. Consumers should not feel shy about asking for local why, because "please explain the decision" is not a request for trade secrets; it's a request for due process in a system that can materially shape your life.

There's also a trap on the consumer side: assuming the black box is always "AI." Sometimes it's not. Sometimes you were denied because of a policy rule: minimum income, minimum time on job, maximum loan-to-value, state restrictions, product capacity limits, fraud concerns that trigger manual review. Those are not model mysteries; they're business choices. A convenient benefit of black-box language is that it lets an institution frame business choices as technical necessities. "The algorithm" becomes a scapegoat.

So when you hear, "The system won't let me," translate it as: "The company has decided to do it this way." That translation restores the possibility of accountability. A company can change a policy. A model can be retrained. A threshold can be adjusted. A mistaken data record can be corrected. Nothing is inevitable. It's just difficult.

The sharpest edge of the black box is that it can launder value judgments as math. "Risk" sounds neutral. But what counts as risky is partly a business preference and partly a moral choice. How much uncertainty is acceptable? Who gets the benefit of the doubt? How much does a past mistake matter? How quickly should forgiveness be priced in? A model will reflect the answers baked into its training data and objectives. When those answers are hidden, the value judgments are hidden too.

You don't need to become a model auditor to protect yourself, but you do need a playbook for when you're on the wrong side of a decision and the explanation feels thin.

The first move is boring and powerful: ask for the adverse action notice (or read it carefully if it's already provided). Don't treat it like junk mail. It's often the only legally anchored explanation you're going to get, and it can point you to the correct lever. If the reasons mention your credit report, pull your reports and scan them for obvious errors. If the reasons mention utilization, look at your balances relative to limits. If the reasons mention length of credit history, you may be dealing with a time problem, not a behavior problem.

The second move is to assume that data is guilty until proven innocent. Not because the world is out to get you, but because databases are messy and identity resolution is hard. If the decision seems inconsistent with your reality, look for mismatched personal information, duplicate accounts, unfamiliar addresses, or accounts that should have been closed. The black box isn't clairvoyant; it's only as good as the data it was fed.

The third move is to build redundancy into your financial life so one black box can't trap you. Keep at least one backup payment method. Maintain a second bank account if you can. Avoid having all your credit availability tied to a single issuer. When a model flags you incorrectly, redundancy turns a crisis into an inconvenience.

All of that is consumer self-defense. But the deeper point is psychological: don't let the black box convince you that you are unworthy. A denial is not a character evaluation. It's a risk estimate under uncertainty, plus a policy threshold, plus a business objective, plus whatever errors are lurking in the data. That's a mouthful, but it's also liberating. It means the decision is not a verdict on you; it's a snapshot from a system that can be wrong and can be changed.

The actionable takeaway is to treat opaque financial decisions like you would treat a surprise medical bill: *ask for the itemization*. In practice, that means you request and save the adverse action notice, you identify whether the decision was driven by credit report data or something else, and you push for specificity until you have at least one lever you can pull within 30 days. If you can't get a clear lever, your lever becomes switching providers: apply elsewhere, use a credit union, choose a different product, or route the decision through a channel with human review. Black boxes thrive when you accept them as final. They weaken when you insist on reasons you can act on.

2.4 Invisible Redlines: Bias and the Data Dragnet

A redline used to be a line you could picture. A banker, a map, a neighborhood outlined in ink, a quiet decision that certain streets were "too risky" for mortgages. It was ugly, concrete, and—importantly—*visible*. You could point to it. You could sue over it. You could write a newspaper story that held it up to the light.

The modern version often doesn't look like a line at all. It looks like a number. A score. A tier. A "risk segment." It arrives with a smiley UX wrapper—*good news, you're pre-approved!*—or with bland corporate regret—*we're unable to offer you coverage at this time.* The boundary is still there, but it's drawn in math, not marker, and the shape shifts depending on what the model thinks you resemble.

This is why the combination of AI and money makes people uneasy even when the marketing promises "fairness" and "objectivity." Algorithms claim neutrality but eat biased data for breakfast. They don't need to know your race, your gender, your disability status, or your religion to treat you differently along those lines. They only need data that *correlates* with those traits—and modern life produces endless correlations.

The problem isn't that data science is inherently discriminatory. The problem is that finance is a mirror, and the mirror has fingerprints. Lending histories reflect decades of unequal access. Insurance data reflects unequal exposure to risk (where you can afford to live, what jobs you can get, how you commute). Employment data reflects discrimination and opportunity gaps. Even *clean* transaction data reflects a society where some people pay more for the same things, have less margin for error, and get punished harder for small setbacks. When a model learns from that past, it can learn the inequities along with the patterns.

Bias in financial AI is often misunderstood as a villain twirling a mustache: a model explicitly using "race" as an input. That's rare and illegal in many contexts. The more common version is quieter and more dangerous: a model using proxy variables—

innocent-seeming bits of information that stand in for protected traits.

A proxy variable is how the system learns what you didn't tell it. You didn't report your race, but you typed a ZIP code. You didn't report your age, but your spending pattern looks like a cohort. You didn't report your health status, but you purchase certain products at certain intervals. You didn't report your immigration status, but your phone number history, address stability, and credit file thickness create a statistical silhouette. The model isn't "thinking" about these identities the way a human bigot would. It's doing something colder: clustering people into predicted outcomes using whatever signals help it reduce error.

From the institution's perspective, that can look like progress. More data can mean more inclusion. Alternative data—information beyond traditional credit history—can help someone with a thin credit file demonstrate reliability. A steady pattern of rent payments might be meaningful. Consistent utility payments might be meaningful. A stable income stream through a bank account might be meaningful even if a borrower has never had a credit card. For consumers who've been invisible to the traditional system, data can be a bridge.

But bridges can also become chokepoints. The same data that expands access can also create new ways to exclude people, because "more data" doesn't automatically mean "better truth." It often means "more surface area for error, inference, and stereotyping."

To see how this happens, imagine a lender trying to predict who will default. The lender has a simple, human-legible model: income, existing debt, payment history. It's imperfect but understandable. Now the lender adds a machine learning model that ingests hundreds of behavioral features: how often your balance hits zero, whether your spending spikes after payday, how quickly you respond to past collection notices, how often you change devices, the time of day you log into the app, the stability of your addresses, the mix of merchants you transact with.

None of those features says "race." None says "gender." But some of them line up with structural realities that differ across groups. If you live in a neighborhood shaped by decades of underinvestment, you may have fewer mainstream banking options, different fee structures, more reliance on prepaid cards, different commuting patterns, more gig work, more income volatility. A model can learn that these patterns correlate with higher default rates—not because the people are morally riskier, but because the environment is harsher and the financial cushion is thinner. The model is not declaring anyone inferior. It is simply optimizing a loss function, and the loss function has no compassion.

This is where people get tripped up: a model can be "accurate" and still be unfair. Accuracy is not the same as justice. If a model predicts that people from certain ZIP codes default more often, it may be correct on average. But pricing or denying based on that correlation can replicate the very conditions that created the default risk in the first place. You don't just measure the world; you reinforce it.

Insurance makes this even sharper because insurance is, in essence, institutionalized prediction. The whole product is priced on risk pooling. When models become more granular—more "personalized"—the pool shrinks. The logic becomes: why should low-risk people subsidize high-risk people? That sounds reasonable until you remember that "risk" isn't always chosen. You don't choose whether your neighborhood has more accidents, or whether your city has more theft, or whether extreme weather is getting worse, or whether your job requires driving at night. Hyper-personalization can turn insurance into a polite way of saying, "You're on your own."

And then there's the other part of the story: the data dragnet.

A dragnet is a fishing method. You cast it wide and pull up everything—fish you want, fish you don't, debris, seaweed, a boot. In modern consumer finance, the dragnet is the way data gets collected because it *might* be useful. App permissions, browsing behavior, device fingerprints, location signals, click paths, social graph hints, third-party data brokers, "partners." Not all companies use all of it, and not all uses are equally invasive. But the economic incentive is always the same: more data might improve prediction, and better prediction is monetizable.

There's a reason this feels creepy even when nothing illegal is happening. The old system asked you for information in a form you could see: income, employer, rent, references. The new system can infer things about you from behavior you didn't realize was being scored. It can take actions you experience as personal—higher interest rates, lower limits, denied coverage—based on signals you didn't know were relevant.

That gap between what you think you're sharing and what's being used to judge you is where trust goes to die.

The dragnet also changes power dynamics in a simple way: it moves the burden of proof onto you. If the model quietly infers you are risky, you now have to prove you are not. But the evidence the model used may be invisible or proprietary. You're being asked to win a game without being told the rules.

Even in cases where firms provide reasons (and in many credit contexts they must), the reasons can be partial or abstract. "Insufficient credit history" might be true, but the model may also be reacting to something more subtle: address instability, income volatility, device changes. Those factors might be proxies for hardship, not irresponsibility. If you respond by taking out a credit-builder loan because you were told you need "history," you might not fix the underlying issue the model is actually reacting to. You spend time and money chasing the wrong lever.

Bias doesn't require bad intentions. That sentence matters because it's the only way to talk honestly about the real world. Bias can enter through at least three doors: the world, the data, and the deployment.

The world door is simple: history is unequal. If past approvals were skewed, then the data about "successful borrowers" is skewed.

The data door is technical: measurement is messy. Errors and missingness are not random. Some groups have thinner files, more name variations, more shared addresses, more cash usage, more informal employment, more phone number churn. A model can interpret missingness as risk. The model doesn't "know" why the data is missing; it just learns that missingness correlates with outcomes.

The deployment door is organizational: a model might be trained responsibly, but then used in a context it wasn't designed for, with thresholds adjusted to hit business targets, with limited monitoring, with poor appeal processes. You can have a decent model and a bad system around it.

And because the dragnet collects *so much* information, the consequences of a mistake can be wide. If your identity is slightly mis-linked—your file tangled with someone else's address history, your device associated with a fraud ring because you bought a used phone, your email flagged because it appears in a breach—those errors can echo across products. A single bad inference can become a multi-industry stigma: banking friction, credit denials, insurance quotes that make no sense, extra verification steps that waste your time.

This brings us to the other risk of the dragnet: cybersecurity.

Data is not just power; it's liability. The more personal data a company pools—especially when it aggregates data from brokers and partners—the more attractive it becomes as a target. A database full of account identifiers, addresses, phone numbers, transaction patterns, and identity signals is a thief's dream. It's also a blackmail dream, a phishing dream, and a synthetic identity dream.

Even when companies protect data well, the mere existence of massive data lakes creates a form of systemic risk. Your information isn't just stored with your bank. It's stored with the vendors your bank uses, the analytics providers, the customer support tools, the marketing platforms, the fraud detection services. Each link is another potential breach point. The dragnet doesn't only gather fish; it gathers hooks.

There's an irony here that's hard to ignore: companies often justify extensive data collection as necessary for safety—fraud prevention, identity verification, money laundering detection. Those are real needs. Yet every extra field collected becomes another field that can be stolen, misused, or misinterpreted. Security and surveillance grow together, like two plants sharing the same soil.

A lot of consumers respond to this by trying to opt out of everything. That's understandable, and sometimes it's the right call. But opting out has costs too. In some systems, less data can mean more friction, more manual verification, lower credit limits, or higher pricing because you're harder to score. The tragedy is that the people who most need access and fair pricing are often the least able to absorb the downsides of privacy-protective choices. Privacy becomes a luxury good.

So what does self-defense look like when the redlines are invisible and the dragnet is everywhere?

It starts with a mindset shift: treat "personalized" as a warning label, not automatically as a benefit. Personalization can mean convenience. It can also mean granular surveillance, granular pricing, and granular exclusion. When a product promises "offers tailored just for you," the hidden translation is: *we have a model of you, and we're going to monetize it.*

It also helps to recognize that fairness in algorithms is not a single switch. You can make a model fairer on one definition and less fair on another. You can equalize approval rates and still produce unequal error rates. You can reduce bias in predictions and still have biased outcomes because the underlying world is unequal. That's not an excuse; it's a reason governance matters. It's also a reason consumers should be skeptical of vague claims like "bias-free AI." There is no such thing as bias-free data about a biased world. There are only choices about which biases you tolerate, which you correct, and which you decide are unacceptable.

Contestability is your friend here too. When a model-driven decision hurts you—denial, higher price, lower limit—your best move is not to argue philosophy. Your best move is to demand actionable reasons and check for correctable errors. In credit decisions, reason codes and adverse action notices are often the key. In insurance, you may have fewer formal explanation rights depending on product and state, but you can still ask what data sources were used and whether there's a consumer report involved (and if so, how to obtain it). When a decision is driven by third-party data, your leverage often runs through the third party.

The most reliable way to reduce proxy-driven harm in your own life is painfully unglamorous: keep your core financial signals clean. Pay on time when you can. Keep credit utilization moderate if you use credit cards. Maintain stable contact information. Monitor your credit reports and dispute errors quickly. This isn't about moral purity; it's about reducing the number of weird statistical shadows you cast. In a dragnet world, clarity is a form of protection.

But you shouldn't have to optimize your life for a model's convenience. The larger solution is policy, oversight, and better institutional behavior: careful testing for disparate impacts, limiting the use of certain alternative data, strong documentation and review, and robust appeal pathways. Consumers can't individually fix systemic bias, but they can understand it well enough to stop blaming themselves for it.

There's one more twist worth mentioning, because it shapes how these systems feel day-to-day: the dragnet doesn't just sort people into "good" and "bad." It can also sort people into "profitable" and "expensive." Some customers are predicted to call support more. Some are predicted to churn quickly. Some are predicted to be price-insensitive. Some are predicted to accept add-on products. That's not a creditworthiness judgment; it's a business judgment. The same machinery used to prevent fraud can be used to optimize revenue extraction. When you feel like you're being herded toward a certain product or fee, you might be.

In that sense, invisible redlines can appear not only as denials but as *steering*. You're not rejected; you're offered the version of the product with worse terms. You're not priced out; you're priced up. You're not blocked; you're slowed down until you accept the alternative. The model doesn't need to say "no" to shape your choices. It can make "yes" expensive.

The actionable takeaway is to run a simple "dragnet audit" on your financial life once a year, and to do it with the same seriousness you'd bring to locking your doors. Pull your credit reports and scan for mismatches. Review app permissions on your phone—especially financial apps—and revoke anything that doesn't make sense (location access is a common one to question). Use strong unique passwords and turn on multifactor authentication for banks, brokerages, and email (email is the skeleton key). And when you face a surprising denial or price jump, assume the model may be reacting to a proxy or an error: ask what data sources were used, request the explanation you're entitled to, and challenge inaccuracies fast. Invisible redlines thrive in silence; the dragnet thrives on passive consent. Your best defense is a little friction in the right places, and a refusal to accept "the algorithm" as the final word.

Chapter 3

The Unchanging Laws of Wealth

Before we plug into the matrix of high-frequency trading and neural networks, we need to stand on solid ground. Imagine AI as a Ferrari: it's powerful and fast, but if you ignore the laws of gravity and friction, you'll just crash faster. In finance, concepts like compound interest, risk-return trade-offs, and diversification are the gravity. No matter how smart the algorithm, it cannot create money out of thin air or eliminate risk without a cost. This chapter explores the unshakeable rules that govern wealth-rules that robots must obey just as strictly as humans.

3.1 Time Value and the Power of Compounding

A weird thing happens when you stop thinking of money as a *number* and start thinking of it as a *machine*. A number just sits there. A machine produces something. And once you see your savings as a machine that can manufacture more savings, you stop obsessing over the tiny knobs—the clever trades, the perfect timing, the "secret" strategy—and start paying attention to the one lever that actually changes your life: *how long the machine gets to run.*

AI is spectacular at knobs. It can scan headlines, parse earnings calls, model patterns in price data, and execute trades in milliseconds. But AI can't do the one thing that turns modest, boring contributions into serious wealth: it can't create *extra decades*. Time is the only input you cannot download, outsource, or optimize after the fact. You either start the clock, or you don't.

Compounding is the name of the mechanism. Plain-English definition: *you earn returns on your earlier returns*. The first year you invest, your money might grow a little. The second year, you're not just growing your original dollars—you're growing last year's growth too. That tiny "plus one more layer" repeats again and again until the growth stops looking linear and starts looking like something alive.

Most people intuitively picture progress like stacking bricks: add a brick each month, and you get a wall. Compounding is more like planting a seed that becomes a tree that drops seeds that become trees. The early years feel almost insulting: you water and water and it looks like nothing is happening. Then, later, the thing takes off and you catch yourself thinking, *Wait. When did it get this big?* That moment is not magic. It's simply time finally showing up on the statement.

There's a reason scams love to say "Don't miss out" and rarely say "Be patient." Patience is not sexy. Patience doesn't screenshot well. Patience doesn't make for great marketing copy. But patience is what makes the math bend in your favor—and the bend is the whole point. In a world where AI can make everything feel faster, compounding is the stubborn reminder that wealth is still mostly grown, not grabbed.

The snowball analogy is popular because it's accurate *and* a little brutal. Picture a snowball rolling downhill. At the top, it's small enough to fit in your palm. Rolling it feels pointless. The hill is cold. Your gloves are wet. A few minutes later, it's big enough that you have to push with your forearms. Near the bottom, it's a boulder that gathers snow just by existing. That's what your portfolio does in the later decades: it gathers money because it's already large. The hardest part is the top of the hill—when you have the least momentum and the most reasons to quit.

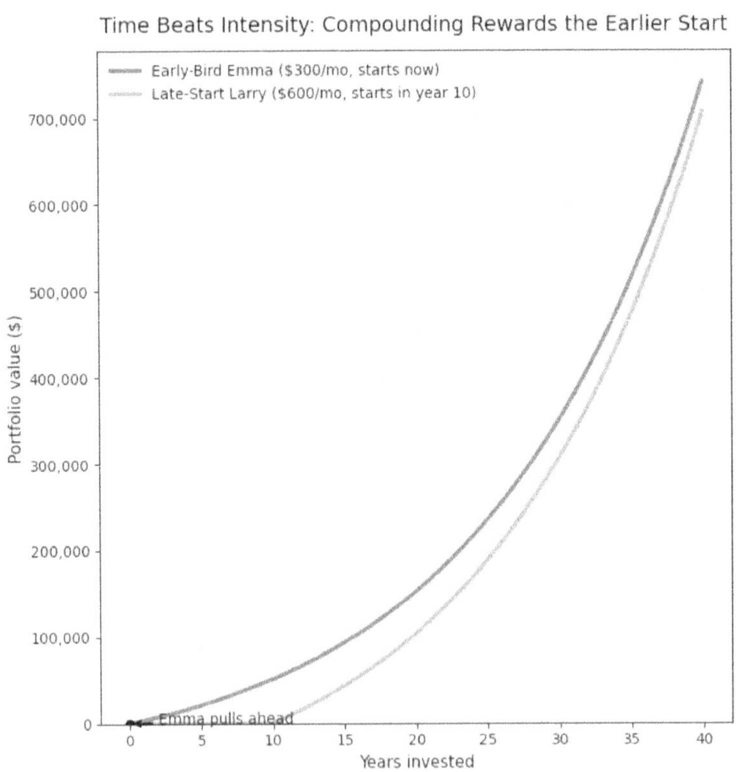

If your eye went straight to the fact that Larry invests twice as much per month, you're thinking like a normal human. Normal humans love effort. We're built to reward visible strain. "He's doing more, so he should win." Compounding is the obnoxious math teacher that says: "Not necessarily." Emma wins in many realistic scenarios because she bought the one asset Larry can never purchase later: the first ten years of growth.

That's the emotional punchline of compounding: you don't just lose the contributions you didn't make; you lose *what those contributions would have become*. Skipping investing at 25 doesn't only mean you have less at 26. It means you have less at 65 *because the missing dollars never had a chance to multiply*. Lost time is not a gap you fill in; it's a branch you never grew.

If you want a quick mental shortcut for why time matters, there's the Rule of 72: divide 72 by your annual return (as a percentage) to estimate how many years it takes money to double. At 8% returns, 72/8 is about 9 years. At 6%, it's about 12 years. This isn't precision engineering; it's a flashlight. It tells you that "a few percentage points" and "a decade" are not small things when doubling is on the table.

Here's the part that people find hard to accept: compounding doesn't care how smart you are *today*. It cares how consistent you are *over time*. A portfolio doesn't ask for your IQ; it asks for your patience. It rewards the person who makes contributions on a boring Tuesday in October, not the person who nails a dramatic trade once.

And this is exactly where AI enters the story in a sneaky way. AI's superpower is reducing friction: auto-investing, smart rebalancing, tax-loss harvesting, personalized nudges, instant account setup. Those are real benefits. But the danger is psychological. When your financial life is wrapped in software that looks like a cockpit, you start believing the cockpit is the point. You forget that the plane still needs distance. A thousand clever features can't compensate for never taking off.

The most expensive financial mistake isn't paying a few bucks too much in fees one year. It's getting spooked, stopping contributions, and sitting out five or ten years because the market feels scary or your feed feels chaotic. When people say "time in the market," they often mean "don't trade a lot." True, but incomplete. The deeper meaning is "don't break the chain." A chain of contributions is more powerful than a chain of predictions.

Compounding has an evil twin, and it's worth meeting. Costs compound too. Fees compound. Taxes compound. Inflation compounds. A small annual drag feels like lint in the moment, and then you wake up later and realize it was a slow leak the whole time. This is why a 1% fee isn't "only 1%." It's a haircut you take every year *on a growing head of hair*. The bigger you get, the more hair gets cut. AI tools that make investing easier are great; AI tools that hide costs behind sleek language are not.

A useful way to think about this is: compounding is a slope, and fees change the steepness of the slope. They don't have to be dramatic to matter. If your net return is 7% instead of 6%, the difference isn't 1% in your life. Over decades, it can be the difference between "comfortable" and "I can breathe." Money is one of the few places where a seemingly tiny number, repeated enough times, becomes a big number without asking permission.

So how do you actually *use* this, beyond nodding politely at the graph?

First, get brutally clear about what your real job is. It's not to find the perfect stock. It's to buy time. That means building a system that keeps you investing even when you're busy, bored, or freaked out. The best system is the one that works when you are not at your best.

That system can be boringly simple:

- *automate contributions*, so you don't rely on motivation;
- *pick an allocation you can live with*, so volatility doesn't force you into panic decisions;
- *keep costs low*, so the compounding engine isn't fighting a constant headwind.

None of those steps require a prediction about next quarter. They require self-knowledge about your future emotions.

Second, treat "starting" as a separate skill from "optimizing." Starting is the highest-return decision because it unlocks years of compounding. Optimizing is fine—later. People love to delay starting until they feel prepared. They research brokerages like they're choosing a spouse. They ask what the best index is. They ask what AI tool is smartest. Meanwhile, the clock is doing what clocks do.

If you're waiting because you don't have much money yet, that's exactly when you should start. Small contributions are not pointless; they are practice reps and time-stamps. They build the habit, and the habit is what keeps you invested when life gets complicated. The first few hundred dollars you invest aren't there to make you rich. They're there to make you *the kind of person who invests*.

Third, understand what "time" really buys you: the ability to be average. Average is underrated. An average investor who contributes consistently for 30 or 40 years can outperform a brilliant but inconsistent investor who starts late, stops often, or overreacts to every headline. AI may raise the average by reducing human error, but it won't repeal the basic arithmetic: the person with more compounding periods usually wins.

There's also a quieter benefit of time that doesn't show up in charts: time reduces the pressure to be right. If you need your portfolio to double in two years, you will be tempted to do desperate things. If you're investing over decades, you can afford to be imperfect. You can survive a bad year, because you have many more years for the math to work. That psychological breathing room is the real luxury money buys. But it starts with time.

One last warning, because the modern investing world is full of shiny buttons: don't confuse motion with progress. AI can generate alerts, scores, watchlists, "conviction" labels, and beautiful dashboards that update every second. Your wealth doesn't grow because your app is busy. It grows because you own productive assets for a long time, and you keep adding to them. The most powerful feature in any investing platform is not a model. It's the setting that moves money from your paycheck into your portfolio on schedule.

Actionable takeaway: set up an automatic monthly investment (even a small one) to run for the next 12 months, then put a recurring calendar reminder for one date each year to increase it by a fixed amount or percentage. Treat that increase like a raise you give your future self. AI can help you click the buttons; only you can keep the clock running.

3.2 Risk, Return, and Volatility

On the day your portfolio drops, your phone becomes a little trap. You open an app for a quick glance, and it hands you a red number like a slap: down $1,842, down 2.7%, down "today." Your brain doesn't experience that as a percentage. It experiences it as a message about you: *you were wrong, you're late, you're behind, you're unsafe.* And then the app helpfully offers buttons that look like relief—sell, "move to safety," "reduce risk," or a bright little suggestion to "switch to a more stable strategy."

This is the moment where investing stops being math and turns into weather. Calm people do fine in sunshine; character shows up in storms. The trouble is that markets are not storms that you can avoid by staying indoors. Markets are *the outdoors.* If you want the long-run returns that markets have historically offered, you have to accept that some days will feel like stepping into cold rain with no umbrella.

Risk is the price of admission. Not as a slogan, but as a structural rule: higher expected return generally requires accepting a higher chance of unpleasant outcomes along the way. If someone offers you high returns *and* low risk as a packaged deal, you should treat it the way you'd treat a stranger offering a free ride to the airport: possible, but the odds aren't in your favor.

Here's where language gets people into trouble. "Risk" sounds like one thing, like a single monster under the bed. In reality, it's a whole zoo. The investing world often uses *volatility* as shorthand for risk, but volatility is only one animal in the enclosure.

Volatility simply means *how much the price bounces around.* It's the day-to-day and month-to-month wobble—the reason your balance can look great in March and bruised in April without anything "fundamental" changing in your life. Volatility is loud, because it comes with headlines. It's also measurable, which is why algorithms love it: you can compute it, chart it, optimize it, and sell products that claim to tame it.

But volatility is not the same as permanent loss. Permanent loss is the kind of damage you can't undo by waiting: a company goes bankrupt, a bond defaults, a fund implodes under leverage, or you sell at the worst possible time and never buy back in. Volatility hurts your feelings. Permanent loss hurts your future.

That distinction matters because a lot of "risk reduction" in consumer investing is really *volatility reduction.* And volatility reduction can be soothing—until it quietly increases other risks you don't notice. A product can look smooth and stable right up until it doesn't. The chart is calm, the marketing is calm, your stomach is calm... and then the thing gates withdrawals, reprices, or reveals that its stability was purchased with hidden leverage and wishful thinking.

If you've ever wondered why the most dangerous financial products often look the *least* scary, it's because humans hate visible pain. We will pay real money to avoid the emotional discomfort of a jagged line. We will sometimes accept much worse dangers if they are dressed up as "steady." This is not a character flaw; it's a wiring diagram.

Volatility is also not a bug in the system. It's a feature of owning uncertain future cash flows. When you buy stocks, you're buying claims on businesses that live in the real world: customers change their minds, competitors show up, regulators rewrite rules, supply chains snap, and brilliant products flop for dumb reasons. The future is messy, so prices are messy. The price moves because the market is constantly trying to update its collective guess about that messy future.

If you want a mental picture, stop imagining a stock price as a "true value" with random noise around it. Imagine it as a live auction for an uncertain future, where the auctioneer never stops talking. Every new piece of information becomes a bid. Most bids are just mood swings. Some are real. Nobody gets to know in advance which is which.

The uncomfortable truth is that volatility is *how the market hands you the possibility of higher returns*. If prices never moved, there would be no discount days. If there were never fear, there would be fewer bargains. If there were never uncertainty, there would be less compensation for taking the leap.

That compensation is what people are reaching for when they invest for the long run. You can call it a risk premium, but you don't need the term to grasp the idea. The market does not pay you extra for doing something easy and certain. It pays you extra for absorbing uncertainty that other people can't stand absorbing.

Now add AI to the story, and the plot thickens.

AI doesn't feel dread. It doesn't get embarrassed at parties because it "bought the top." It doesn't wake up at 3 a.m. and decide to dump everything because a financial influencer posted a scary chart. That sounds like an advantage—and it is, in narrow ways.

The bigger change is that AI can *package* volatility to make it feel smaller than it is. A robo-advisor can show you a neat risk score. A trading app can show you a clean "confidence" meter. A fintech can smooth the chart by updating less often, or by highlighting a longer time frame, or by automatically sweeping losses into a tidy category called "temporary drawdown."

None of those are lies. They're interface choices. But interface choices shape behavior. The more a product tries to hide the sensation of risk, the more likely you are to accidentally buy something you don't truly understand—or to abandon a sound plan the first time the sensation leaks through.

There's a reason casinos have no clocks. It's not because clocks are immoral. It's because time makes people act differently. In investing, volatility is the clock. It reminds you that things are changing. Some platforms try to make you forget that. Others shove your face in it. Neither extreme is great.

The healthiest relationship with volatility is not denial. It's *expectation management*. You want a plan that can survive a bad year without requiring you to become a different person.

That means you need to decide, in advance, what kind of pain you're willing to feel for what kind of potential outcome.

This is where the phrase "risk tolerance" becomes real. Risk tolerance is not what you say when markets are calm. It's what you do when you're down and your brain starts writing dramatic fiction about your future.

A practical way to translate this into real life is to think in two clocks instead of one.

Clock one is your goal clock: when you actually need the money. Down payment in two years? Retirement in thirty? Tuition next fall? Time horizon isn't just a detail; it's the main ingredient in what kinds of risk you can afford to take. Money you need soon has no time to heal. Money you won't touch for decades can endure a lot of drama.

Clock two is your emotional clock: how often you are going to look. This sounds almost silly until you notice the obvious: the more frequently you check, the more volatility you experience. A long-term investor who checks once a year will see a mostly upward story with occasional chapters of pain. A long-term investor who checks five times a day will see chaos, because daily data is mostly noise with occasional meaning.

This is not an argument for ignorance; it's an argument for choosing a viewing schedule that matches your temperament. AI thrives on engagement. Many financial products get paid when you interact, trade, or upgrade. So the incentives often point toward more checking, more alerts, more "insights." Your incentives should point toward fewer, better decisions.

Risk also changes shape depending on what you own. Stocks can drop sharply, but they also can recover strongly because businesses can grow. Bonds are often presented as "safe," but they carry their own set of risks that can be quieter and more confusing: interest-rate risk (rates rise, bond prices fall), credit risk (the borrower can't pay), inflation risk (your fixed payments buy less over time), liquidity risk (harder to sell quickly without a discount), and call risk (the issuer pays you back early when it's convenient for *them*). None of these risks require a dramatic red candle on a chart to hurt you.

This is why "low volatility" isn't the same as "low risk." A bond fund can look calm and still deliver a disappointing decade if inflation runs hot or rates rise. A cash-like product can feel stable and still quietly lose purchasing power every month. The pain is just stealthier.

So what should a normal person conclude from all this, besides "everything is risky"?

Start with a more honest scoreboard.

The investing world often encourages you to judge yourself on short-term outcomes: Did you beat the market this quarter? Did you buy before the rally? Did you avoid the dip? That scoreboard is addictive, but it's also unfair. It confuses luck with skill and it trains you to chase feelings instead of building wealth.

A better scoreboard is: Did you stick to your plan? Did you keep costs reasonable? Did you avoid catastrophic mistakes? Did you stay invested through the unpleasant parts? These are boring questions, which is exactly why they work. Boring questions create boring habits. Boring habits create real money.

You also want to stop treating volatility as a personal insult. A down market is not a verdict on your intelligence. It's the normal process of pricing uncertainty. The market does not know you exist. This can be depressing, but it's also freeing. If the market isn't targeting you, you don't have to take it personally.

What about diversification—the one finance concept that sounds like advice your grandmother would give and yet sits at the foundation of modern investing?

Diversification means spreading your money across different investments so you aren't hostage to a single outcome. It's often called the only "free lunch" in finance because, done well, it can reduce the bumpiness of your ride without necessarily reducing expected long-run return. The intuition is simple: if your eggs don't all break the same way, you get to keep more omelets.

The less obvious part is *why* it works. Diversification isn't about owning a lot of things for the sake of it. It's about owning things that don't all move together. If every holding in your portfolio is basically the same bet dressed in different logos, you haven't diversified; you've just created the illusion of safety with more line items.

AI can help here in a legitimate way. Algorithms are good at measuring how assets have moved relative to one another in the past and at keeping your portfolio near a target mix through rebalancing (selling a bit of what's up, buying a bit of what's down). Rebalancing feels counterintuitive because it asks you to trim winners and add to losers—which is emotionally difficult and mechanically easy. It's the kind of task that software can do without ego.

But even diversification has limits that people forget precisely when they most need to remember them. In big market crashes, many risky assets become more correlated—they fall together because fear becomes the dominant force. Diversification can reduce some risks, but it can't erase the basic truth that owning risky assets means sometimes experiencing risky outcomes. If a tool promises to remove that, the tool is either wrong or redefining "risk" in a way that should worry you.

And now, the most useful reframing: volatility is the *rent* you pay to be in the wealth-building neighborhood.

You don't pay rent because you love paying rent. You pay rent because you want access to a location. Market volatility is similar. You don't endure it because it's fun. You endure it because it's the cost of owning assets that can grow over time. The mistake is trying to avoid rent *and* still live downtown.

There's a second mistake that's just as common: paying rent on a place you can't stand living in. That's what happens when someone holds a portfolio that is too aggressive for their temperament. They panic-sell during downturns, then sit in cash waiting for "clarity," then buy back in higher because it finally "feels safe." The math isn't the problem; the behavior is. And behavior is shaped by how much volatility you can emotionally tolerate.

This is why "appropriate risk" is personal, not moral. Conservative isn't cowardly. Aggressive isn't enlightened. The best portfolio is the one you can hold. The second-best portfolio is the one you can hold after making it slightly less exciting. The worst portfolio is the one you abandon at the wrong moment because it turned you into someone you don't like.

AI-driven products sometimes make this harder by turning risk into a game. They add confetti for gains and quiet little sympathy messages for losses. They show you leaderboards. They suggest "trending" assets. They send notifications that feel like a friend tapping you on the shoulder. All of that can be fun. It can also pull you away from the one habit that makes risk worth taking: staying the course.

If you want a simple test of whether an investing tool is helping you or hurting you, ask what it encourages on a bad day. Does it nudge you toward discipline—or toward action? Disciplined inactivity is not passive. It's a choice. And it's often the most profitable choice you'll make.

Actionable takeaway: write down a one-paragraph "panic plan" and save it where you will actually see it (notes app, printed sheet, whatever works). Include

- the maximum portfolio drop you believe you can tolerate without selling, stated plainly (e.g., "I expect I could see a 30% decline at some point"),
- a rule for what you will do during a downturn (e.g., "I will keep my automatic contributions running and I will not sell based on headlines"), and
- the one date you'll review your portfolio on purpose (e.g., the first Saturday of January).

When the red numbers arrive, you don't need a new prediction—you need your own prior decision.

3.3 Asset Classes: The Building Blocks

A robo-advisor can feel like a magician. You answer a few questions—age, goal, maybe how you'd react if your account fell 20%—and it spits out a crisp portfolio with a name that sounds like a yoga pose: *Balanced Growth, Moderate, Aggressive.* Tap once and you're invested. Tap twice and you're diversified. Tap three times and you're apparently a responsible adult.

Under the hood, though, the magic trick is almost always the same. No matter how futuristic the interface looks, no matter how many graphs it animates, the portfolio is built from a small pantry of ingredients. Learn those ingredients and you'll never again be impressed by a menu. You'll read it like someone who knows how the kitchen works.

Those ingredients are called *asset classes*: broad categories of investments that behave differently because they're tied to different economic promises. The three you'll meet in almost every long-term portfolio are:

- *Stocks*: ownership in companies.
- *Bonds*: loans to governments or companies.
- *Cash*: money you can use quickly, with minimal price fluctuation.

That's it. Everything else is usually a remix, a wrapper, or a fancier label stuck on one of those three.

This matters because the modern financial world loves to sell you the wrapper. "AI-powered." "Smart." "Dynamic." "Risk-managed." "Alternative." "Buffered." "Enhanced income."

These phrases are not automatically bad. But they often distract from the only question that really controls what happens to your money: *What do I actually own, and why does it make (or lose) money?*

Think of your portfolio like a recipe. You can plate it on elegant dishes, drizzle it with a sauce called "machine learning," and sprinkle on something marketed as "institutional-grade." If it's still mostly flour, sugar, and butter, it's still a cookie. The label doesn't change the ingredients; the ingredients change the outcome.

Stocks: owning the upside (and the stomachaches)

When you buy a stock, you're buying a slice of a business. Not a souvenir certificate. A real claim on future profits. If the business sells more, earns more, and becomes more valuable, your slice can become more valuable too. If the business stumbles, your slice shrinks. It's not poetic; it's contractual.

This is why stocks have historically been the main engine of long-run growth in many portfolios. Businesses can grow in a way that fixed promises can't: they can raise prices, invent new products, expand into new markets, and get more efficient. In a world where inflation exists and technology keeps rewriting what's possible, the ability to grow is a superpower.

But you don't get the upside without the attitude problem. Stocks are moody. Their prices can drop fast and for reasons that feel unfair. Sometimes they fall because the business is genuinely deteriorating. Sometimes they fall because investors are suddenly nervous about interest rates, or a recession, or a war, or because everyone decided last month's enthusiasm was embarrassing.

Owning stocks means agreeing to live with that. It's not an academic statement; it's an emotional one. You are signing up for days when the market votes "thumbs down" on your retirement plan for no personal reason at all.

There's another detail people forget until it matters: in the capital structure of a company, stockholders are last in line if things go badly. If a company goes bankrupt, bondholders and other creditors generally get paid before shareholders. Stock is the purest form of "ownership," which also means it's the purest form of "you eat the losses first."

That sounds grim. It's also why stock returns can be higher over long periods: the market compensates you for being the person willing to sit in the back of the line and still show up.

AI-driven investing often changes how people *interact* with stocks, not what stocks *are*. Algorithms can screen for value, momentum, quality, or any number of factors. They can execute trades efficiently. They can even write plausible narratives about why a stock is attractive, because they're trained on oceans of language about markets.

But AI can't make ownership risk disappear. It can only move it around, slice it thinner, or hide it in a product that doesn't look like a stock but still behaves like one.

A simple example: a "low-volatility equity strategy" is still equity. It may bounce less than the overall stock market, but it's not a magic carpet that floats above recessions. In a real panic, it will feel like a stock portfolio, because it is one.

If you want the grown-up version of stock investing, it's this: you're buying participation in human ingenuity and human chaos at the same time. The returns come from the ingenuity. The volatility comes from the chaos. You don't get to unsubscribe from one while keeping the other.

Bonds: lending money (and learning what ``safe'' really means)

Bonds are often introduced as the sensible sibling of stocks: the calm one, the grown-up one, the one that doesn't slam doors. A bond is a loan. You give an issuer money, and the issuer promises to pay interest and return your principal on a schedule.

Because a bond has promised payments, it can feel reassuring in a way stocks don't. A stock can't promise you anything. A bond, at least on paper, can. That's why bonds are the backbone of many portfolios designed for stability, income, or shorter time horizons.

But bonds have a secret: the word "safe" does a lot of lying in bond world.

Some bonds are genuinely high quality, especially those backed by strong governments or solid companies. Others are shaky. And even high-quality bonds can deliver unpleasant surprises if you don't understand the main types of bond risk.

The risk people *notice* least at first is interest-rate risk. Bond prices move *opposite* interest rates. When new bonds are being issued at higher rates, older bonds paying lower rates become less attractive, so their market prices fall. This can be deeply confusing if you think "bonds are safe" means "bond funds never go down."

Bond funds can go down. They can go down for long stretches when rates rise. That's not a scandal; it's the normal math of fixed payments being repriced.

Then there's credit risk: the borrower might not pay. That's why U.S. Treasury bonds are generally treated differently from bonds issued by a shaky company. The promise is only as good as the promiser.

Inflation risk is the quieter cousin. If a bond pays you 4% and inflation runs at 4%, you're treading water in purchasing power. If inflation runs at 6%, you're losing ground, even if the bond pays every penny on time. The bond did what it promised. The economy changed the meaning of the promise.

Liquidity risk and call risk are the kinds of things that sound like fine print until you experience them. Liquidity risk means it may be hard to sell a bond quickly at a fair price. Call risk means the issuer may pay you back early, usually when interest rates fall—exactly when you would have liked to keep earning the higher rate you locked in. It's like a tenant who can cancel the lease whenever it benefits them, not you.

So why hold bonds at all?

Because bonds can serve jobs that stocks are bad at. Bonds can dampen portfolio swings. They can provide a more predictable stream of income. They can be a source of funds to rebalance from during stock market downturns. They can match liabilities that have a specific date (tuition, a home down payment, a planned retirement date). Bonds are the portfolio's shock absorbers and calendar tools.

AI can add value in bond investing in a very unglamorous way: by helping with implementation. Building a bond ladder, selecting maturities, managing duration, reinvesting coupons, and maintaining a target allocation are tasks that software can do consistently. Consistency is underrated because it isn't a headline, but it is a real advantage.

What AI can't do is repeal the basic bond tradeoff: if you want higher yield, you generally accept more credit risk, more interest-rate sensitivity, or both. A higher yield is not a free gift. It is a billboard advertising that the bond carries more worry.

Cash: the underrated superpower of optionality

Cash is the ingredient people love to insult. Cash is called lazy. Cash is mocked as "doing nothing." Cash is accused of "missing out." And for long-term wealth building, cash *can* be a drag if you hoard it instead of investing.

But cash has a superpower that doesn't show up in return charts: *optionality*. Cash gives you the ability to act without selling something at a bad time. Cash pays you in flexibility.

The most important cash in your financial life is not "dry powder" for investing. It's the money that keeps you from becoming a forced seller. Emergency funds. Near-term bills. The cushion that lets you keep investing through a layoff, a surprise medical bill, or a car that decides to die on the way to work.

When people say "cash is trash," what they often mean is that inflation erodes purchasing power. That's true. Cash is not a long-term growth asset. It's a short-term stability asset. Its job is not to win the race; it's to keep you from falling off the track.

There's also a subtle point that gets missed in the AI era: cash reduces the chance that an algorithm can corner you emotionally. If your app is constantly nudging you to optimize and trade and shift, cash is the part of your plan that says, "We're not desperate." Desperation is expensive. It leads to bad trades, bad timing, and bad products.

Cash in a portfolio is not a confession that you don't believe in markets. It's an admission that you live in a life where surprises happen, and you'd like to survive them without detonating your long-term plan.

Cash has its own risks too, of course. The biggest is inflation risk: the money is stable in dollars, but dollars may buy less over time. There's also opportunity cost: holding too much for too long can mean you don't participate in long-run growth. Like every other asset class, cash is a tool. Used well, it saves you. Used poorly, it stalls you.

The wrappers: funds and ETFs (the container is not the food)

Now for the part that confuses almost everyone at first: many of the things you buy are not asset classes. They're wrappers.

A mutual fund or an ETF (exchange-traded fund) is a container that holds investments. The fund pools money from many investors and buys a basket of underlying assets: stocks, bonds, sometimes cash-like instruments. Your return comes from what those underlying holdings do, plus any distributions (like dividends or capital gains) the fund passes through, minus the fund's costs.

This is why "I own an ETF" is like saying "I ate a bowl." What was in the bowl?

An ETF can be 100% U.S. stocks. It can be long-term government bonds. It can be short-term corporate bonds. It can be a blend. The wrapper doesn't tell you the risk. The ingredients do.

AI sometimes makes this harder because it can label wrappers in comforting ways. A fund can be marketed as "income" when the income is coming from taking on extra credit risk. It can be marketed as "buffered" when the buffer is paid for by giving up some upside. It can be marketed as "enhanced" when the enhancement is simply leverage or complexity.

Complexity isn't evil. But complexity is often expensive, and expensive is a problem because costs compound in the wrong direction. If a product is complicated, ask a simple question: *Which ingredient is doing the work? Stocks, bonds, or cash? And what new risk did the product add to make the marketing claim true?*

How the ingredients work together: a portfolio is a job assignment

The point of learning asset classes isn't to win trivia night. It's to assign jobs.

Stocks are your long-term growth engine. Bonds are your stability and scheduling tool. Cash is your flexibility and survival tool. A portfolio is not a single bet; it's a small organization where each asset class has a role.

When someone says, "My portfolio is 80/20," they're describing staffing. Eighty percent in a growth role, twenty percent in a stabilizing role. When someone says, "I moved to all cash," they fired the growth team and hired only security guards. Sometimes that's appropriate for money you need soon. Sometimes it's a panic move that feels good today and costs you later.

The AI twist here is that modern platforms can make the staffing feel automatic—set a risk level, get a mix, rebalanced for you. That's genuinely convenient. The danger is that convenience can turn into mindlessness. If you don't know what each ingredient does, you can't tell whether the platform is helping you build a sturdy plan or simply selling you an attractive story.

A useful habit is to translate any portfolio back into the three ingredients. If you can't explain your own mix in plain language, you're probably renting complexity you don't need.

To make this practical, here's a simple table you can use as a decoder ring when you're staring at a product description that feels like it was written by a committee of robots.

Ingredient	*What it really is*	*What can go wrong (in plain language)*
Stocks	Owning businesses (claims on future profits)	Prices can drop sharply; individual companies can fail; headlines can whipsaw sentiment
Bonds	Lending money (scheduled interest + principal)	Rates can rise (prices fall); borrower can default; inflation can erode what payments are worth
Cash	Liquidity and stability in dollar terms	Inflation can quietly shrink purchasing power; holding too much can slow long-term growth

The table is simple on purpose. Real portfolios can include real estate, commodities, private assets, and more. But even many of those ultimately behave like combinations of the big three: ownership-like risk, lending-like risk, or cash-like stability. If you master the basic ingredients, the rest becomes easier to evaluate without getting hypnotized.

And here's the final, slightly uncomfortable truth: most investing stress comes from a mismatch between what people *think* they bought and what they actually bought.

They think they bought "steady income," but they bought risky credit. They think they bought "safety," but they bought long-term bonds sensitive to rates. They think they bought "AI-managed protection," but they bought stocks with a fancier dashboard.

Actionable takeaway: pull up your portfolio holdings and force them into three buckets—stocks, bonds, cash (you can add "other" if you must, but don't let "other" become a junk drawer). Write down, in one sentence per bucket, the job it serves for you. If you can't name the job, reduce or remove the holding until you can. The goal isn't to simplify your life for aesthetic reasons; it's to make sure your money is built from ingredients you recognize, so no algorithm or marketing slogan can sell you a recipe you didn't mean to eat.

3.4 Indexing vs. Active Management

The most seductive sentence in investing is also the shortest: *"I can do better."* It shows up in quiet ways—a coworker bragging about a stock pick, an app that flashes "top movers," a podcast host who sounds uncannily certain. It also shows up in sophisticated ways: factor models, alternative data, machine-learning forecasts, sentiment analysis, satellite imagery of parking lots. The costumes change. The impulse doesn't.

Indexing and active management are two different answers to the same question: *When you buy investments, are you trying to capture what the market gives everyone, or are you trying to extract something extra?*

Indexing says: "Give me the market return, as cheaply and cleanly as possible." Active management says: "I'm going to beat the market, or protect you from it, by making different choices than the market." One approach treats the market like a weather system you dress for. The other treats it like a puzzle you can solve.

Both can be intelligent. Both can be misused. But they don't start on equal footing, and the reason has nothing to do with IQ.

The market isn't an opponent. It's a crowd.

Imagine the stock market as a giant stadium where everyone is buying and selling the same set of slices of future profits. Some people are day traders. Some are retirees. Some are pension funds. Some are algorithms that never blink. Every time you buy a stock, someone else sells it to you. Every time you sell, someone else buys.

That simple fact creates a structural constraint that's easy to miss: *before costs, the average active investor must, by definition, earn the market return*. They are collectively *the market*. If one active manager beats the market, another must trail it by roughly the same amount, because they're trading with each other. This is why people call it a zero-sum game *before* costs.

Then costs show up, and the game becomes negative-sum for the group.

Costs aren't just the explicit fee you see in a prospectus. They include trading costs (spreads, market impact), taxes (in taxable accounts), and the subtle friction of being wrong at the wrong time. After those costs, the average active dollar tends to underperform the market. Not because markets are holy, but because arithmetic is rude.

Indexing is, at its core, a decision to stop paying for the privilege of trying to be above average.

What an index fund really is (and what it isn't)

An index is a list with rules. The S&P 500 is a list with rules. A total market index is a list with rules. A bond index is a list with rules. The index describes *what* is included and *how much* of each thing you own (often by market value).

An index fund (mutual fund or ETF) is a product designed to track that list. It does not mean "no decisions were made." The decisions are simply moved earlier in the chain: someone chose the index, someone defined the rules, and someone chose how the fund will track it (own every security, sample a subset, sometimes use derivatives). Indexing is not the absence of judgment; it's the decision to buy a transparent rulebook instead of a manager's hunches.

That transparency is part of the appeal. With indexing, you usually know what you're buying. With active management, you often buy a story about what the manager *might* do.

Index funds also come with a small but important reality check: you can't invest directly in an index. You invest in a fund that aims to track it, which creates the possibility of *tracking error*—the fund doesn't perfectly match the index due to costs, cash drag, sampling, and the mechanics of real-world trading. Indexing is simple; it is not literal.

Still, the philosophy is powerful: take the market's return, minimize the leakage, and let time do the heavy lifting.

Active management: the two promises (and the third one they don't say out loud)

Active management usually sells one of two promises:

- *Outperformance*: beat the benchmark.
- *Risk management*: lose less in downturns, or smooth the ride.

There's a third promise, rarely said outright because it sounds less noble: *emotional comfort*. Many active products exist because investors crave the feeling that someone is "at the wheel." In a scary market, paying a professional to make decisions can feel like hiring a pilot for turbulence. Sometimes that's worth something. Sometimes it's just expensive reassurance.

The hardest part is that these promises can all be true in the short run. A manager can beat the market for a year, three years, even five. A manager can also protect on the way down for a while. The question isn't "can it happen?" The question is "can it happen reliably enough, net of fees, to justify choosing it over a cheaper alternative?"

That reliability is rare. Not impossible. Rare.

And rare is a problem because the average investor is shopping for rare skill in a market that is excellent at manufacturing convincing narratives.

Why fees are a bigger deal than your brain wants to admit

If a fund charges 1% per year and a comparable index fund charges 0.05%, the difference feels like pocket change. It's not. It's a permanent headwind.

Fees don't just reduce your returns once. They reduce the amount of money that can compound for you year after year. That compounding of *what you didn't get to keep* is why costs matter so much more than they feel like they should.

Active management begins every year in a small hole: it must beat the market by at least its additional costs just to tie. That's a high bar even for genuine talent, because the market is not a lazy opponent.

Indexing doesn't guarantee you'll get rich. It does something more modest and more reliable: it prevents you from making a common mistake—overpaying for the hope of outperformance.

The most honest sentence about AI investing

AI has changed how investing feels. It has not changed what investing *is*. The market is still a crowd. Stocks are still ownership. Bonds are still loans. Risk still demands compensation.

So where does AI fit into indexing versus active management?

Most "AI investing" is simply active management with a new engine.

If an AI model is selecting stocks, timing entries, predicting returns, rotating sectors, trading quickly, or responding dynamically to news, that's active management. It might be brilliant active management. It might be misguided active management. But it's active, because it is making *different* choices than the market portfolio in pursuit of an advantage.

This isn't a criticism. It's a classification.

The important consumer move is to stop letting "AI" act like a moral upgrade. AI does not turn a difficult game into an easy one. It may improve execution. It may discover patterns. It may reduce certain kinds of human error. But it also introduces new risks: model risk (the pattern stops working), data risk (the inputs are biased or stale), regime risk (the world changes), and crowding risk (too many algorithms chase the same signal, and the advantage evaporates).

AI can also make active management more tempting because it makes it feel less like gambling. A person with a hunch feels like a person with a hunch. A model with a dashboard feels like science. Sometimes it is. Sometimes it's just math-shaped confidence.

If you want one sentence to keep you grounded: *AI can change the method; it doesn't change the hurdle.* The hurdle is still: can you outperform after costs, consistently, in a market full of competitors who also have models?

The haystack and the needle (and why most people should buy the haystack)

The "needle in a haystack" metaphor persists because it's emotionally accurate. Beating the market feels like finding something special. It feels like discovery. Indexing feels like settling.

But the market isn't a single haystack. It's a haystack where millions of very smart, very motivated people are already looking for needles, and they get paid based on whether they find them. Many are supported by research teams, data feeds, and computing power that dwarf anything a household investor can justify.

In that environment, buying the haystack isn't laziness. It's a strategy that says: "If needles exist, I will own them automatically, in proportion to their importance, without having to predict which ones they are."

Indexing turns the market's collective intelligence into something you can rent cheaply.

Active management asks you to believe you can hire or become someone who can consistently do better than that collective intelligence *after* costs.

That doesn't mean active management is always wrong. It means it should be treated as a luxury purchase, not a default.

When active management can be rational (not just hopeful)

Some active choices can make sense, but notice how different these reasons are from the usual marketing:

You have a specific constraint the index ignores. For example: a desire to avoid certain industries, a need to match cash flows, a tax situation that benefits from customization. Some of that can be achieved with indexed building blocks; some may require active decisions.

You're buying a form of risk management you truly understand. Not the vague promise of "downside protection," but a clearly defined approach with tradeoffs you can live with—for example, accepting capped upside in exchange for a buffer, or holding higher-quality bonds even if yields are lower.

You have strong reasons to believe a manager has durable edge. This is the hardest one. Durable edge is rare, and even when it exists, it's often capacity constrained (it doesn't scale). Past performance is a noisy signal. Marketing materials are not evidence. The burden of proof should be high because the cost of being wrong is decades of compounding lost to fees and underperformance.

The most common bad reason to go active is also the most human: *boredom.* Indexing can feel anticlimactic. It doesn't give you stories to tell at dinner. It doesn't give you the dopamine hit of being right. If you need entertainment, buy entertainment. Don't turn your retirement plan into a personality.

A practical way to interrogate any AI investing product

When an app says "AI-managed portfolio" or "AI stock picks," ask two questions that cut through the fog:

- *What benchmark should I compare this to?* If there is no clear benchmark, it's hard to know whether you're paying for skill or paying for vibes.
- *Where, exactly, is the return supposed to come from?* Cheaper implementation? Better diversification? Faster trading? Better forecasts? Reduced taxes? If the answer is a handwave about "finding opportunities," assume you are being sold hope.

Then ask the question that most people avoid because it feels impolite: *What does it cost, in all-in terms?* Advisory fees, fund expenses, transaction costs, spreads, platform fees. If the product can't give you a clear estimate, that's information too.

The point isn't cynicism. It's self-defense. The market is competitive; the investing product market is even more competitive, and it competes for your attention as much as your returns.

The behavioral advantage of indexing

One of indexing's biggest benefits isn't mathematical. It's psychological.

Indexing reduces the number of decisions you can regret. It lowers the temptation to chase performance. It removes the feeling that you should "do something" every time a headline hits. It turns investing from a game you can lose through overactivity into a system that mostly runs on rails.

If you're using AI tools, this matters even more. AI tools are excellent at presenting you with *actionable* information: signals, scores, alerts, rankings. The existence of a signal creates the feeling that action is required. Indexing is a deliberate refusal to be yanked around by the stream of "actionable" noise.

You can still use AI in an indexed approach, in a way that's genuinely helpful: automating contributions, rebalancing, tax-aware asset location, and preventing obvious mistakes. That's not AI trying to outsmart the market. That's AI helping you execute a sensible plan with fewer errors.

Actionable takeaway: choose a default stance for your money that you can explain in one sentence, and force every "AI" pitch to argue against it. A strong default for many people is: *"My core portfolio will be broad, low-cost index funds; any active or AI strategy must be small, clearly benchmarked, and worth its all-in cost."* Then write down the exact percentage you'll allow for experiments (even if it's 0%). The goal isn't to ban curiosity; it's to stop curiosity from quietly becoming your entire financial plan.

Chapter 4

Your Pilot is an Algorithm

Imagine walking into a prestigious Wall Street firm with $500 in your pocket. Ten years ago, you would have been politely shown the door. Today, an algorithm rolls out the red carpet. But who-or what-is actually flying the plane with your retirement savings? We're going to lift the hood on the 'robo-advisor' to reveal a machine that is less about sci-fi magic and more about ruthless efficiency. We will look at how it translates your anxieties into asset allocation, why it sells your winners when you want to keep them, and how it turns market crashes into tax breaks-all without you lifting a finger.

4.1 Inside the Robo-Advisor: From Questionnaire to Portfolio

Picture this: it's a Sunday night, your laundry is half-folded, and you finally open a robo-advisor app because you promised yourself you'd "get serious" about investing. The app is calm, almost soothing. No shouting stock tickers. No adrenaline. Just a handful of questions that feel suspiciously like a personality quiz: your age, your income, when you might need the money, how you'd react if the market fell 20%, whether you've invested before.

Ten minutes later, you're handed a portfolio that looks like it came out of a Wall Street research department: a global mix of U.S. stocks, international stocks, emerging markets, bonds, maybe real estate. Neatly priced. Cleanly labeled. Delivered with the confidence of a pilot announcing, "We'll be cruising at thirty-seven thousand feet."

That jump—from a short questionnaire to a disciplined portfolio—is the whole magic trick. And the trick is mostly not magic. It's translation.

A robo-advisor is best understood as a machine that turns messy human life into a small set of numbers, then uses those numbers to select from a limited menu of portfolios built from low-cost funds. The algorithm isn't "predicting the next Apple." It's doing something far more practical and, honestly, more useful: choosing *how much risk you're going to take* and enforcing that choice consistently, even when your emotions beg for exceptions.

The reason it can feel like a black box is that the most important step happens in the middle: *your answers get compressed*. The app can't ingest the full reality of your life—your job stability, your family obligations, your sleep quality during a market crash, the fact that you might want to start a business, the fact that your spouse gets hives when money is discussed. So it squeezes all of that into a few inputs that are easy to score and easy to map to a portfolio.

That compression is both the superpower and the weakness. A robo-advisor can offer surprisingly good investing defaults at scale because it reduces the world to a small set of decisions it can execute cheaply. But if the questionnaire misses something important, the portfolio can be "optimized" for a person you resemble only on paper.

The questionnaire isn't asking about risk. It's asking about regret.

Most risk questionnaires sound like they're measuring your willingness to take volatility. In practice, many of them are measuring something more human: how likely you are to panic and sabotage yourself at the worst possible moment.

Consider the classic prompt: "If your portfolio dropped 20% in a year, what would you do?" The technically correct answer might be "buy more," but the emotionally honest answer might be "stare at the ceiling at 3 a.m., then sell everything by lunchtime." Robo-advisors can't monitor your stress hormones, so they use scenario questions as a proxy for your *behavioral breaking point*.

The uncomfortable truth: the biggest investing risk for many people isn't that markets will be volatile. It's that *you will be volatile*. A portfolio that is mathematically "optimal" but psychologically intolerable is not optimal at all. It's a beautifully engineered car that you refuse to drive because the brakes squeal and you're convinced something is wrong.

So the questionnaire typically tries to estimate three things:

- *Capacity for risk*: Can you afford downturns without ruining your plan? (Time horizon, stable income, emergency fund, other assets.)
- *Need for risk*: Do you need higher returns to reach your goal, or will a safer path do? (Goal size, savings rate, timeline.)

- *Tolerance for risk*: Will you stick with the plan when it stops feeling good? (Behavior under stress.)

The app may not label them that way. But that's what it's sniffing for. The output is usually a risk score or a risk "band"—conservative, moderate, aggressive—that will later snap into a model portfolio.

And here's where the human drama sneaks in: if you answer aspirationally instead of honestly, you can end up with a portfolio that is less "aggressive" than it is *fragile*. Fragile means it works right up until the moment you abandon it.

The upside of the questionnaire is that it forces a decision most people avoid: picking a risk level *before* the market tests their courage. The downside is that it can feel like a few questions are allowed to decide your financial personality.

A funnel that turns life into a pie chart

Right after the questionnaire, the robo-advisor typically shows you the destination: a target allocation. Often it looks like a pie chart split between stocks and bonds. Sometimes it's broken into more slices: U.S. stocks, developed international stocks, emerging markets, U.S. bonds, international bonds, inflation-protected bonds, and so on.

That portfolio may look bespoke. In many cases, it's a selection from a small set of pre-built models. The personalization is real, but it's often *categorical*. You might be placed into "Portfolio 7 of 10." That's not an insult; it's how scale works. You're not getting a custom-built engine. You're getting a well-tested engine in the size that fits your car.

Here's the translation pipeline in its simplest form: your inputs enter at the top, the algorithm processes them through a set of investing rules, and a portfolio comes out the bottom.

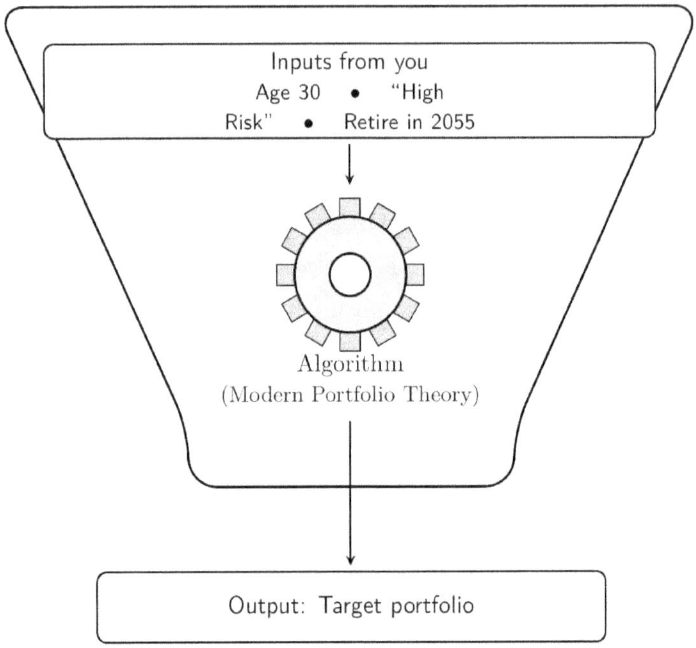

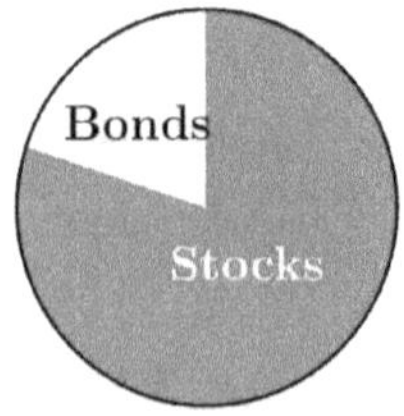

Example allocation: *80% Stocks / 20% Bonds*

The important thing to notice about that funnel is what *isn't* inside it: market forecasts. A robo-advisor generally doesn't need to predict which sector will win next year to be useful. It needs to do three unglamorous jobs well: choose a risk level, diversify broadly, and keep costs and mistakes low.

Modern Portfolio Theory: the boring math behind the shiny interface

Many robo-advisors lean on a framework called Modern Portfolio Theory (MPT). The name sounds like a TED Talk from 1964, which is not far off: it's a set of ideas about how to combine assets so the whole portfolio has an attractive balance of risk and return.

In plain English, MPT says: don't judge investments one by one. Judge how they behave *together*. If two assets tend to zig and zag at different times, mixing them can reduce the stomach-lurching swings without necessarily sacrificing expected return as much as you'd think.

That's the origin story of the basic stocks-and-bonds mix. Stocks are volatile but historically have offered higher long-term returns. High-quality bonds tend to be steadier and, in some bad stock markets, have cushioned the fall. The mix aims to give you a smoother ride.

But MPT is also where the "black box" feeling can intensify, because it relies on inputs that look scientific and are easy to misunderstand:

- *Expected return*: a guess about long-run performance (never guaranteed, always model-dependent).

- *Volatility*: how violently an asset price tends to swing.
- *Correlation*: how two assets move relative to each other.

These are not eternal constants like gravity. They can change. Correlations that looked comforting in one decade can become less helpful in another. That doesn't make MPT useless—it makes it a *tool*, not a prophecy.

A robo-advisor's advantage is not that it has secret knowledge about the future. Its advantage is that it can apply a consistent philosophy to millions of accounts without charging each person for a team of humans.

Why your ``custom" portfolio looks like ETFs

Once you're assigned a target allocation, the robo-advisor has to implement it. This is where reality—and cost—forces discipline.

Most robos build portfolios primarily using low-cost index funds, usually ETFs (exchange-traded funds). An ETF is a basket you can buy and sell like a stock. Many ETFs track an index: the S&P 500, a total U.S. stock market index, a total bond market index, an international stock index, and so on.

Why ETFs, specifically?

Because they're cheap in exactly the way that matters. If you buy a handful of broad ETFs, you can get diversification across thousands of companies and bonds while paying very low annual expenses. Those expenses might look tiny—0.03% here, 0.10% there—but investing is a game of compounding. Costs compound too, except in the wrong direction.

ETFs also fit the operational needs of automation. They're liquid, easy to trade, easy to rebalance, and easy to scale. If you're running an algorithm that must maintain a target mix for hundreds of thousands of clients, you want instruments that behave predictably when you buy and sell them.

There's also a quieter, more human reason: ETFs make the portfolio less dramatic. A robo-advisor is often trying to keep you from doing something rash. A portfolio of a few broad funds doesn't tempt you to "just tweak" it every time a headline appears. It's harder to fall in love with a total international index fund. That's a feature, not a bug.

None of this means ETFs are the only good investment vehicle, or that active management is always bad. It means that for an automated service promising broad diversification at low cost, ETFs are the natural building blocks.

The menu behind the curtain

Even when the interface shows you a clean allocation, there's usually a menu behind it. The algorithm often chooses from a limited set of model portfolios that have been designed, tested, and approved internally. Think of it like ordering a burrito: you pick your protein and your spice level, but you're not designing the supply chain.

A platform might have, say, ten risk levels. Each level corresponds to a different stock/bond mix and sometimes different slices within stocks and bonds. Your risk score determines which model you get.

That sounds less personal than people expect, but it solves a real problem: consistency. If an automated platform let every account become a bespoke snowflake, it would be harder to monitor, explain, and control. Regulators, auditors, and risk managers tend to like systems that can be described as rules rather than vibes.

The point is not that you're being shortchanged. The point is that robo-advising is *mass customization*. It's the financial equivalent of prescription lenses made by a machine: not hand-crafted, but dramatically better than squinting.

The hidden assumptions you're agreeing to (without realizing it)

When you accept a robo-advisor's recommendation, you're not just picking a pie chart. You're accepting a bundle of assumptions that are rarely displayed with the pie chart itself.

Here are the big ones that matter in real life:

Assumption 1: Your timeline is real. If you say you won't touch the money until 2055, the algorithm will treat that as a law of physics. But humans revise timelines all the time: a job loss, a move, a divorce, a health issue, a dream opportunity. A portfolio designed for "don't touch for 30 years" can feel terrifying when life changes and the timeline quietly becomes "maybe five."

Assumption 2: Your risk tolerance is stable. The questionnaire captures how you *think* you'll react. Markets test how you *actually* react. A robo can't prevent panic selling if you override it, but it can reduce the chance you end up in a portfolio that practically begs you to panic.

Assumption 3: Diversification will behave as expected. Diversification is not a magic shield; it's a risk-management technique. In some market regimes, bonds hedge stocks nicely. In others, they don't. A model portfolio is built on history and theory, neither of which signs a contract with the future.

Assumption 4: Costs and taxes matter more than cleverness. Most robos are implicitly saying: the reliable edge available to normal investors is not stock picking. It's keeping fees low, staying diversified, and minimizing behavioral mistakes. That's not cynical. It's an attempt to be honest about what tends to be repeatable.

If you disagree with those assumptions, the portfolio might still be fine—but you should know you're disagreeing, not just clicking "continue."

A quick walk-through with a real-feeling investor

Meet Maya. She's 30. She has a steady job, $25,000 in cash savings, and she's finally ready to invest. She tells the robo-advisor she's investing for retirement, likely around 2055. She checks a box that says she can tolerate "high risk", because she's heard that young people should be aggressive, and she wants the highest long-term return.

The algorithm doesn't picture Maya as a person. It pictures her as a tuple of variables:

- long time horizon

- no near-term withdrawals (supposedly)
- high willingness to stomach volatility

That tuple maps to a high-equity model, something like 80% stocks and 20% bonds.

Then comes the ETF selection. Within “stocks,” the robo might split across U.S. and international markets. It might sprinkle in emerging markets. Within “bonds,” it might pick high-quality government and investment-grade corporate bonds, possibly with some inflation protection.

Maya doesn’t need to know the underlying tickers to benefit from the structure. But she does need to know what the structure implies: if markets drop sharply next year, her account will likely drop sharply too. That isn’t a bug. That’s what 80% stocks means.

Now imagine the same Maya, same age, same timeline, but she answers one question differently: “If your portfolio dropped 20%, what would you do?” She selects “sell some to reduce risk,” because she knows herself. The portfolio might come back as 60% stocks and 40% bonds.

In a bull market, that second portfolio may feel boring. In a bear market, it may feel like the difference between staying invested and quitting. *The best portfolio is the one you can hold* is not a motivational poster. It’s a mathematical statement about compounding under real human constraints.

The robo's real promise: fewer unforced errors

The most underrated achievement of robo-advisors is that they take a pile of good investing habits that used to require effort and package them as defaults. They make it easier to:

- start with diversification instead of a handful of familiar stocks,
- keep fees from silently draining returns,
- avoid turning every headline into a portfolio change.

This is not trivial. Many investors don't underperform because they picked the wrong index fund. They underperform because they change their plan too often, or because they never had a plan to begin with.

That said, automation can also make mistakes feel more alien. When a human advisor recommends a portfolio, you can interrogate them. You can watch their face when you ask, "What happens if I lose my job?" An algorithm won't blush. It will simply continue.

So the healthiest way to use a robo-advisor is not to treat it as a clairvoyant. Treat it as a *guardrail*. It can keep you from wandering into common ditches. It cannot know where you're going if you don't tell it the truth about your destination.

Actionable takeaway: re-take the questionnaire like a pessimist

Once a year—and anytime your life changes materially—revisit your robo-advisor questionnaire and answer it in the mood you're most likely to have during a bad market day.

A practical method: when you're about to submit "high risk", pause and ask, *If my portfolio dropped 30% between now and December 31, 2026, would I still sleep? Would I still keep contributing?* If the honest answer is no, choose the portfolio you can stick with, not the one that impresses an imaginary finance professor. Your future self doesn't need bravado; your future self needs consistency.

4.2 The Myth and Math of Personalization

Personalization has become the word you say when you want something to feel expensive. Personalized sneakers. Personalized skincare. Personalized playlists that somehow know youŕe sad before you do. In investing, personalization gets marketed like a superpower: *Finally, a portfolio as unique as you are.*

And then you open the hood and discover the uncomfortable truth: for years, "personalized" often meant you were placed into one of five pies.

That isnt́ necessarily bad. A small menu of well-built portfolios can beat the pants off a random collection of hot tips. The myth is that personalization must be intricate to be valuable. The math is that a little personalization goes a long way—and after that, it gets complicated fast.

The new wrinkle is that software has made a deeper kind of customization possible without hiring a private banker: *direct indexing*. Instead of buying an ETF that tracks the S&P 500, you can own the S&P 500's ingredients (or a close approximation) directly, then edit the recipe: remove the companies you don't want, tilt toward the ones you do, and potentially squeeze more tax efficiency out of the same market exposure.

It's a bigger lever. It also comes with bigger ways to hurt yourself.

The first era: ``Choose your risk level." The rest is the same sandwich.

Early robo-advisors were built like a very competent cafeteria. You chose your tray—conservative, balanced, aggressive—and the kitchen served the same meal to thousands of people who made the same choice. It was mass production with a friendly interface.

If that sounds cynical, try a different frame: it was industrial-grade consistency. Most new investors don't need 200 decisions. They need one decision that sticks: *how much of my money should be in risky stuff versus stabilizers?* The minute you do that, you've done more "financial planning" than most people manage in a decade.

So the first wave of personalization was mostly about risk bands. Age, time horizon, and a few behavioral questions would nudge you into a stock/bond mix. From there, the portfolios were generally built from broad ETFs, because the goal wasn't to express your identity. The goal was to get you diversified, invested, and paying low fees.

The myth crept in when marketing language started implying that the portfolio was handcrafted for your life. In reality, your answers were often a key that unlocked one of a few model portfolios.

Again: not necessarily a scam. But itś important to understand the difference between *personalization as a feeling* and *personalization as a change in what you actually own.*

Personalization is expensive---not in dollars, but in complexity

Every additional dimension of personalization is another thing that can go wrong. Thatś not a moral judgment; itś physics.

If a platform customizes your portfolio based on your tax bracket, it must know your tax bracket and keep it updated. If it customizes around your outside assets—your spouseś 401(k), your company stock, the rental property youŕe emotionally attached to—it has to collect that information, interpret it correctly, and update it when your life inevitably changes.

Most people doń update their beneficiaries, let alone their full balance sheet.

So platforms make trade-offs. They personalize what they can measure reliably and implement cheaply. That tends to mean: risk level, time horizon, account type (taxable versus retirement), and sometimes simple preferences like "ESG tilt" or "avoid fossil fuels."

The danger is not that the algorithm is evil. The danger is that the investor assumes the algorithm knows things it has never been told.

A personalized portfolio built on stale or incomplete inputs is like a tailored suit measured while youŕe wearing a winter coat. It fits beautifully until you take the coat off.

The second era: ``You can have the index... but with edits."

Direct indexing changes the *unit* of investing. Instead of owning one ETF share that represents a slice of 500 companies, you own many individual stocks (or a representative subset), each with its own cost basis and tax lot history.

That last phrase—*tax lot*—sounds like something youd́ avoid at a party, but itś where the power comes from. A tax lot is simply a record of when you bought a particular set of shares and what you paid for them. When you own a single ETF, you have fewer lots, fewer moving parts. When you own hundreds of individual stocks, you have a garden full of lots, each one potentially useful when you want to realize gains or losses strategically.

This is why direct indexing used to be mostly for wealthy investors with teams: itś administratively annoying. Software made it administratively possible.

And once you can own the index as ingredients, you can do three kinds of personalization that are materially different from choosing "Portfolio 7 of 10."

Values-based exclusions. Dont́ want tobacco? Remove it. Dont́ want oil and gas? Remove it. Dont́ want a particular company because you work there and your paycheck already depends on it? Remove it.

This is the cleanest "because I said so" form of personalization. It feels empowering, and it is. But it's also the most likely to be underestimated, because the cost isn't a line-item fee. The cost is *tracking error*—your results drifting away from the index you think you're tracking.

Factor tilts. Instead of simply mirroring the index, you can tilt toward characteristics like value (cheaper companies), quality (more profitable, less debt), momentum (recent winners), small size, or low volatility.

This can be sophisticated, and it can also be a fancy way of saying, "I want to overweight what recently looked good." The math here isn't the problem. The problem is patience. Factor tilts can underperform for long stretches, which is when human beings tend to abandon them right before they work again.

Tax personalization. Owning the ingredients makes it easier to harvest losses across many positions, not just at the fund level. In a volatile market, a direct-indexing strategy can sometimes find more opportunities to realize losses without dramatically changing your market exposure.

This is the most practical form of personalization because it has a real-world payoff: after-tax returns. But it also depends on your tax situation, your account type, and how well the system avoids creating unwanted capital gains while chasing losses.

Direct indexing, in other words, can make the portfolio feel more *you*. It can also make it more *fragile*, because there are more knobs to turn—and more ways to turn them at the wrong time.

The psychological benefit is real: people invest better when they feel respected

There's a slightly awkward truth about investing: the part that looks like math is often compliance with your own emotions.

If an investor hates owning a certain industry and feels guilty every time they see it in their holdings, they will be tempted to tinker. They will hover over their account like a person checking a text thread that isn't going well. Direct indexing can reduce that psychological friction. It lets people say, "This portfolio reflects my life and my values," which can make them more likely to stick with it.

That can be worth something even if the portfolio is not perfectly "optimal" in an academic sense. A plan you adhere to beats a plan you abandon.

But there's a twist: the same customization that increases commitment can increase entitlement. Once you start editing the index, it's easy to believe you're doing more than expressing values. You start believing you're improving the index. You start believing the market will reward your preferences.

Markets do not care about your preferences.

Tracking error: the polite name for ``your portfolio did something different"

Tracking error is a technical term that describes how much your returns deviate from a benchmark. In everyday language: itś the gap between what happened to *the thing you meant to own* and what happened to *the thing you actually own*.

Tracking error isnt́ inherently bad. If you exclude a sector you strongly dislike, you should expect to drift from the index. Thatś the point. The problem is that tracking error has an emotional cost that shows up at the worst possible moment.

Imagine you choose a "S&P 500, minus fossil fuels" direct-index strategy. The market has a year where energy stocks soar. Your friends brag about "the market" being up 18%, and your account is up 14%. Nothing is broken. You got exactly what you asked for: a portfolio that doesnt́ fully track the benchmark. But the feeling is brutal: *Iḿ losing to the market.*

This is where investors often commit a quiet act of self-betrayal. They abandon the customization when it underperforms, then re-adopt it when it becomes popular again. They turn values into a timing strategy. Thatś a recipe for chronic dissatisfaction.

The math of tracking error is straightforward. The psychology is not.

A useful mental model is to treat customization like ordering food with substitutions. If you ask for "no cheese," you dont́ get to be shocked that it tastes different. The key is deciding whether you want the substitution *even when it makes the meal less delicious in some moments.*

``Delete the companies you hate" sounds simple until you ask: which companies are those, exactly?

One reason direct indexing can get messy is that values are not a checkbox; theyŕe an argument.

Take "fossil fuels." Do you mean oil producers? Pipeline companies? Utilities that burn coal? Banks that finance drilling? Industrials that manufacture components? Tech companies whose data centers consume enormous energy? The supply chain is a hydra: cut off one head, and you discover three more.

So platforms often use predefined screens: they rely on third-party data providers that categorize companies. Thatś not wrong, but it means your "personal values" are often implemented through someone elseś taxonomy. Two providers can disagree about whether a company belongs on the naughty list. Your portfolio can shift because of a reclassification, not because you changed your mind.

This matters because personalization is frequently sold as control. In practice, itś often *delegated control*, with a user interface that makes it feel direct.

If youŕe going to customize, itś worth understanding whether the platform is doing:

- *simple exclusions* (remove certain tickers or categories),
- *threshold-based screens* (remove companies above some revenue percentage from an activity),
- *scoring/tilting* (underweight "bad" scores rather than fully excluding),

- *proxy substitution* (replace excluded names with similar ones to keep sector exposure).

Those are different choices, and they lead to different kinds of tracking error. The interface may not highlight the difference, because the difference is not aesthetically pleasing. But itś financially meaningful.

Direct indexing also changes how you experience ownership

An ETF is psychologically smooth. You own "the market." Itś abstract. A direct-index portfolio can feel like owning a roster.

You will see names you recognize and names you dont́. You may find yourself forming opinions about individual holdings you never would have noticed inside an ETF wrapper. You may start checking earnings headlines for companies that represent 0.12% of your account, because now they have a name and a logo.

That sounds harmless. It can become a trap. The more individual stocks you see, the easier it is to slip into stock-picking brain. You start asking whether you should remove a company because it "seems risky," not because it violates a rule you set ahead of time.

Customization invites meddling. Meddling invites second-guessing. Second-guessing invites turnover. Turnover can invite taxes and mistakes.

If the goal is long-term, low-drama compounding, the best investing interface is often the one that makes you slightly bored.

The math trade-off: concentration sneaks in through the back door

Thereś a less obvious mathematical cost to exclusions: removing a set of companies doesnt́ just remove their moral baggage. It changes your portfolioś diversification.

If you exclude enough names, your portfolio can become unintentionally concentrated in certain sectors or styles. You might think youŕe simply "removing harmful companies," but youŕe also removing exposure to certain parts of the economy. Your portfolio may become more sensitive to specific risks (interest rates, commodity shocks, regulatory changes, currency swings).

Good direct-indexing systems try to manage this by using optimization: they aim to keep the portfolioś overall characteristics close to the benchmark while honoring constraints. But constraints have a price. The more constraints you add, the harder it becomes to mirror the index.

Thereś also a practical issue: if you have a smaller account balance, you canť perfectly replicate an index with hundreds of names without running into trading frictions (like minimum share sizes, liquidity considerations, and the sheer number of positions). Many solutions use sampling owning a representative subset rather than all 500 names. That can work well, but it adds another layer of approximation.

So "direct indexing" is not a binary. Itś a spectrum:

- full replication (own nearly all constituents),
- sampled replication (own many, but not all),

- custom factor/value overlays (own a modified version by design).

Each step away from the benchmark is either a feature you intentionally choose or a cost you unintentionally pay. The problem is that the interface often treats all deviation as sophistication, when deviation is simply deviation.

The personalization you actually need is usually boring

For most people, the highest-impact personalization isn't deleting five companies. It's aligning the portfolio with realities that don't fit neatly into a risk quiz:

- You have a big emergency fund need because your income is volatile.
- You hold a huge amount of employer stock and already have concentrated risk.
- You're saving for a home down payment in three years, so "high risk" is a terrible fit for that bucket of money.
- You have significant assets in a spouse's retirement plan that create an overall allocation you should account for.

Those are personalization questions that can change your life. They're also annoying to collect and hard to automate. Which means they're often underemphasized compared to the customization that looks good in a product demo.

This is where the myth becomes financially dangerous: believing you're "fully personalized" because your portfolio aligns with your values, while it quietly fails to align with your cash needs, your job risk, and your true timeline.

A portfolio can be ethically aligned and financially misaligned at the same time.

So when is direct indexing worth it? When the trade actually matters.

Direct indexing can be a smart tool when you have a clear reason that goes beyond novelty.

One reason is tax sensitivity: if you're investing a meaningful amount in a taxable account, and you're in a situation where harvesting losses and managing gains could plausibly improve after-tax outcomes, owning more ingredients gives the system more opportunities.

Another reason is strong, stable preferences: if you have exclusions you care about deeply enough to live with underperformance in some years, direct indexing can help you commit without constantly renegotiating with yourself.

A weaker reason is vague discomfort: "I kind of don't like some companies." That's not nothing, but it's not a stable constraint. If it wobbles every time the market changes, it can turn customization into churn.

Direct indexing isn't magic. It's leverage. Leverage amplifies your intent. If your intent is coherent, leverage can be helpful. If your intent is mushy, leverage can turn mush into a mess.

Actionable takeaway: write your personalization rules while the market is quiet

Before you click any customization toggle, write down two sentences and keep them somewhere you'll see again:

1. *My goal for personalization is:* (values alignment, tax efficiency, avoiding employer concentration, etc.)
2. *I will stick with these rules even if it underperforms the benchmark for:* (one year, three years, a full market cycle).

If you can't fill in the second sentence without squirming, you don't have a personalization strategy yet. You have a mood. And moods make expensive portfolio decisions right before they evaporate.

4.3 The Discipline Machine: Automated Rebalancing

The cruelest part of investing is that the *right* move often feels like a bad personality trait.

Selling something that just made you money feels disloyal, like walking out of a party right when it gets fun. Buying something that just lost money feels naive, like volunteering to carry a piano up the stairs. The market doesn't reward feelings, but it does know how to press on them. It will hand you a winner and whisper, *Don't you dare sell this.* It will hand you a loser and hiss, *Why would you add more?*

Rebalancing is the practice of doing both anyway.

If robo-advisors deserve their keep, it's not because they can pick tomorrow's winners. It's because they can perform this small act of financial self-control over and over, without needing a pep talk, a podcast, or a glass of wine. They can be, in the most literal sense, a discipline machine.

What rebalancing actually is (and why it's so emotionally weird)

A rebalanced portfolio is one that keeps returning to its intended mix. If your target is 60% stocks and 40% bonds, you don't just set that once and walk away. Markets move. Stocks sprint. Bonds shuffle. Over time, your mix drifts.

Rebalancing means trimming what has grown too large and adding to what has shrunk too small, so your portfolio stays aligned with the risk level you chose when you were calm and rational.

It sounds tidy. It is not tidy in your head.

The emotional weirdness comes from the fact that rebalancing is a form of *contrarianism with a seatbelt*. You are systematically selling the asset that has been rewarding you and buying the asset that has been disappointing you. It's "sell high, buy low," not as a slogan, but as a maintenance routine.

And maintenance is precisely the right vibe. Nobody expects their car to run forever on the oil it shipped with. But many investors expect their portfolio to maintain its intended risk profile without ever being adjusted. That's magical thinking, and the market is happy to charge you tuition for it.

Drift: when your "safe" portfolio quietly becomes a different animal

Drift is what happens when the market rewrites your risk profile without asking permission.

Say you start with $100,000: $60,000 in stocks and $40,000 in bonds. Then stocks have a great run. Bonds do fine, but not spectacular. A year or two later, you might look up and discover you're sitting at 75/25 or 80/20.

Nothing dramatic happened in a single day. No one broke into your account and flipped a switch labeled *make it spicy*. Drift did it gradually. Drift is stealthy like that.

The problem isn't that 80/20 is immoral. The problem is that it may be a portfolio you never agreed to hold.

When the next downturn arrives—and downturns are not optional in equity investing—you're not just riding out a market decline. You're riding out a decline with *more exposure to the thing that's falling* than you intended. That's how people end up shocked by losses that were, in retrospect, completely predictable.

Drift also creates a psychological trap: the best time to rebalance is often after your risky assets have done well, which is exactly when you feel smartest and least interested in changing anything. You start to believe your portfolio is working because you are clever, not because markets have a way of making most strategies look brilliant during the right stretch of weather.

Rebalancing is the moment you stop mistaking luck for identity.

The simplest math: why a target mix is really a risk thermostat

A target allocation isn't primarily about return. It's about *how much pain you're signing up for* in a bad market.

Stocks have historically offered higher long-term returns than high-quality bonds, but they arrive with larger and more frequent drawdowns (big drops from peak to trough). A 60/40 mix is not "moderately optimistic." It's a thermostat setting. It aims to keep your portfolio within a band of volatility you can tolerate while still participating meaningfully in growth.

If drift pushes you toward 80/20, it's like your thermostat quietly sliding from 70 degrees to 83 because the sun came out. You'll notice when the heat wave arrives.

A human can rebalance. But humans tend to rebalance late, rarely, or not at all, because it involves doing something that feels like sabotaging success. Algorithms, blessedly, do not experience sabotaging success. They experience an if-statement.

Where the algorithm's ``values" live: bands, triggers, and taxes

If you want to know what a robo-advisor truly believes about investing, don't read its marketing. Find out how it rebalances.

Rebalancing isn't one universal rule. It's a family of choices:

- *How far is "too far" from target?* Some systems rebalance when an asset class drifts beyond a narrow band (say, 2–3 percentage points). Others allow wider drift (5 points or more) before trading.
- *How often is the portfolio checked?* Many platforms monitor daily but trade only when thresholds are hit.
- *What counts as a trigger?* Deposits, withdrawals, dividends, and big market moves can all cause trades. Some

systems use new cash flows first to reduce the need to sell.

- *How tax-sensitive is it?* In taxable accounts, selling an appreciated position can create capital gains taxes. A tax-aware algorithm may tolerate more drift, or preferentially rebalance using contributions and dividends, because taxes are friction.

That last bullet is the subtle one. Rebalancing is often framed as purely a risk-control activity, but in taxable accounts it becomes a negotiation between risk and taxes. You might *want* to sell stocks to get back to 60/40, but selling could generate a tax bill. A well-designed system tries to get you closer to target without creating unnecessary taxes.

This is where "automation" stops being a convenience and becomes an engineering problem. Two platforms can both claim "automatic rebalancing" and deliver very different real-world behavior.

A picture of drift (and the snap-back moment)

It helps to see drift the way an algorithm sees it: as a line wandering away from a target, then getting tugged back when it crosses a boundary.

Below is a stylized example using a 60/40 target. The drifting portfolio gradually becomes more stock-heavy in a rising market, creeping toward 80/20. The rebalanced portfolio is allowed to drift a bit, but when it hits a threshold the algorithm "snaps" it back toward 60/40.

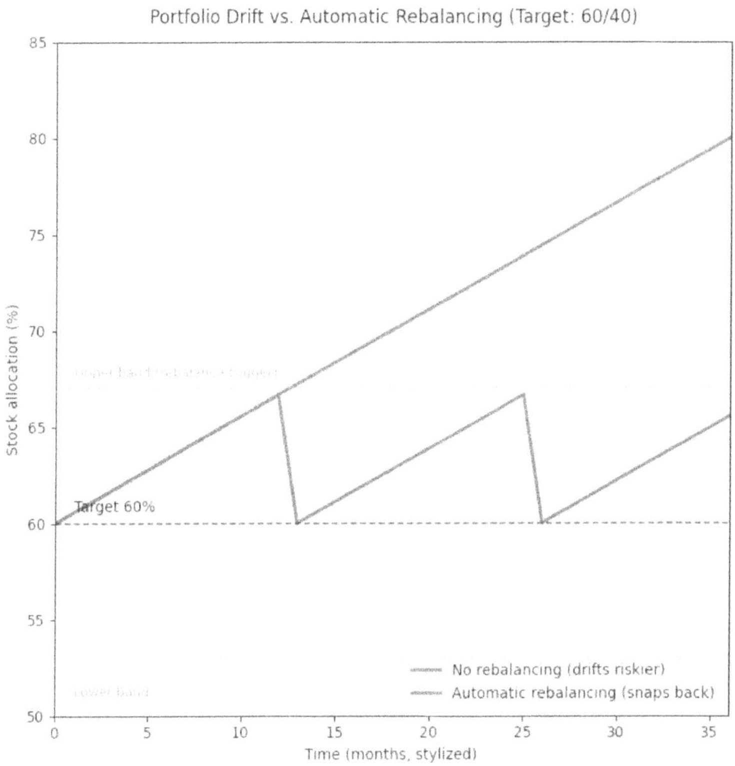

The chart is intentionally simple, because the core idea is simple: if you don't rebalance, your portfolio becomes a different portfolio. If you do rebalance, you keep re-choosing your original risk level, even when the market tempts you to "let it ride."

``Sell high, buy low" is not a prediction. It's a reset button.

Rebalancing is sometimes discussed like a return-enhancement strategy. People talk about a "rebalancing bonus," the idea that regularly trimming winners and adding to laggards can add a bit of performance over time, especially in choppy markets.

That bonus can exist. It can also vanish. It's not guaranteed, and it's not the point.

The point is that rebalancing is a *risk control* practice first. It prevents your portfolio from becoming accidentally concentrated. It keeps you from waking up one day and realizing your retirement plan turned into a high-volatility bet because the last few years were good.

In that sense, "sell high, buy low" is less about being clever and more about being consistent. It's a reset button you press regardless of the news cycle.

This is also why rebalancing can feel infuriating in real time. When the market is roaring, rebalancing can look like you're clipping your own wings. When the market is falling, rebalancing can look like you're throwing good money after bad.

Both interpretations miss the point: you're not trying to win the next quarter. You're trying to keep your plan intact.

The human failure mode: rebalancing only after panic

A lot of investors "rebalance" in a way that resembles a fire drill. They ignore drift in good times, then become intensely interested in risk control right after the market punishes them.

That is the worst timing.

If you wait to reduce risk until after a large decline, you're often selling depressed assets. If you wait to add risk until after a large rally, you're often buying elevated assets. Your emotions will reliably encourage you to do the opposite of what a disciplined policy would do.

This is where automation earns its keep. The algorithm can trade when you're busy, bored, or asleep. It can do it without needing you to feel brave.

Even better: it can be boring. Boring is underrated. Boring is what you want from a system whose job is to keep you from turning investing into a series of dramatic personal statements.

Rebalancing in the real world: contributions, dividends, and ``do the least harmful thing"

A smart robo-advisor doesn't just rebalance by selling winners. Selling can be costly in taxable accounts, and it can be unnecessary if you have ongoing cash flow.

So many systems try to rebalance by *steering new money*. When your paycheck contribution arrives, the algorithm can direct it toward the underweight asset classes. When dividends come in, it can use them the same way. This is rebalancing without as much selling, which can mean fewer taxes and lower trading costs.

Think of it like turning a ship with the engines rather than slamming the rudder. You still change direction, but you do it in a way that wastes less fuel.

Withdrawals create the reverse problem: if you're taking money out, the algorithm might sell from overweight positions to keep the remaining portfolio closer to target. The goal isn't perfection; it's keeping your risk exposure from drifting wildly due to routine life events.

This practical, slightly unglamorous "do the least harmful thing" approach is where automated advice often beats DIY in the real world. People are busy. People forget. Algorithms don't.

A warning label worth taking seriously: automation will rebalance even when you are emotionally unprepared

There's a flip side to discipline: it can surprise you.

If you didn't understand what you signed up for, you might be startled to see the system selling a fund that has been doing great. You might feel the urge to turn the feature off. Or worse, you might override trades in a way that defeats the whole point.

So it's worth being clear-eyed: automatic rebalancing is not a promise to maximize returns. It is a promise to maintain a chosen risk profile through mechanical actions you might not personally choose in the moment.

If that sounds too rigid, it might be because you're imagining rebalancing as a judgment about which asset is "good." It isn't. It's a judgment about *how much of each asset you want to own*, regardless of recent performance.

The algorithm is not praising bonds when it buys them. It is not condemning stocks when it sells them. It is obeying your prior decision about risk.

Actionable takeaway: set your rebalancing expectations before you notice the trades

Open your robo-advisor settings (or disclosures) and find three specifics: (1) whether rebalancing is threshold-based (bands) or calendar-based, (2) whether the system uses cash flows (contributions/dividends) to rebalance before selling, and (3) how it handles taxes in taxable accounts.

Then do one more thing that sounds almost too simple: write down your target stock/bond mix (like 60/40) in a note on your phone. The next time the market has a great year and your portfolio quietly drifts riskier, you'll have a number to compare against—a reminder of what you chose *before* the market started flattering you.

4.4 Turning Lemons into Tax Credits: Automated Loss Harvesting

The most irritating thing about a market downturn is how personal it feels. You didn't *choose* for your international stock fund to drop. You didn't *invite* interest rates to jump and push bond prices down. Yet there it is, in your account, sitting like a dent in a new car.

Tax-loss harvesting is the rare investing tactic that looks at that dent and says: fine. Let's at least get the insurance payout.

It doesn't "beat the market" in the way headlines mean it. What it can do is turn some of the pain into a tax asset—a deduction you can use to reduce future tax bills. When it works well, it's not a flashy trick. It's plumbing. It's the kind of boring optimization that feels almost unfair once you understand it, because it mostly rewards people who were already patient enough to stay invested when prices were falling.

And that's why software is so good at it. Humans hate interacting with losses. Algorithms don't. They see losses the way a mechanic sees a worn brake pad: a maintenance event with a procedure.

A loss isn't useful until you make it real

There's a difference between a loss you *feel* and a loss the tax code recognizes.

If an investment drops in value but you keep holding it, you have an *unrealized loss*. It's real to your net worth, but it's not yet real to your taxes. The IRS doesn't grade you on how bad your portfolio chart looks. It cares about what you *realize*—what you actually sell.

Tax-loss harvesting (often shortened to TLH) is the practice of selling an investment that's down, so the loss becomes *realized*. Once realized, that loss can generally be used to offset realized capital gains. And if you have more losses than gains, you may be able to use a limited amount to offset ordinary income each year, with the rest carried forward to future years (subject to tax rules).

That one step—turning unrealized into realized—is where the strategy lives. It is also where most people freeze. Selling a losing investment feels like admitting you were wrong, even if you were "wrong" about something as ordinary as the market having a bad quarter.

Software doesn't need to protect its ego. It can press the button.

The core move: sell the lemon, buy a similar lemon immediately

The part that makes TLH feel like a cheat code is that, done properly, it doesn't require you to abandon your investing plan.

The mechanical version looks like this:

- You own Fund A. It drops. You're sitting on a loss.
- The algorithm sells Fund A, locking in the loss for tax purposes.
- The algorithm immediately buys Fund B—a substitute that aims to keep you exposed to the same part of the market.

So you stay invested, but you also bank a loss.

This is where robo-advisors and automated platforms shine. They can scan accounts continuously, find opportunities across many holdings, execute trades quickly, and do the recordkeeping without you building a spreadsheet that becomes your new part-time job.

It also pairs naturally with the rebalancing logic you've already seen: both are rule-based trading systems. Rebalancing is about controlling risk drift. TLH is about improving after-tax outcomes by using volatility as raw material.

The market drops, and a human sees only damage. The algorithm sees both damage and deductible.

The wash sale rule: the trapdoor under the whole strategy

Here's the catch, and it's not optional: you don't get to sell for a tax loss and then immediately buy the same thing back.

The wash sale rule is the tax system's way of saying, "Nice try."

In general terms, if you sell a security at a loss and buy the same or a *substantially identical* security within a window around that sale, the IRS can disallow the loss. The window is 30 days before and 30 days after the sale date—a 61-day span that makes TLH feel like it comes with an invisible tripwire.

If the loss is disallowed, it doesn't necessarily vanish into the void. Often it gets added to the cost basis of the replacement shares, effectively postponing the loss rather than erasing it. But postponing is not the same as claiming. If the whole point was to reduce taxes this year, "postponed indefinitely" is not a satisfying outcome.

This is why TLH platforms are obsessed with substitutions. They need Fund B to be similar enough that you stay invested, but not so similar that it's deemed substantially identical. The tax code does not provide a perfectly crisp, universally loved definition of "substantially identical" for all the modern fund combinations investors actually use, which adds a layer of judgment and conservatism to platform designs.

So TLH is less like "sell and buy back" and more like a supermarket switch: you put the name-brand cereal back on the shelf and grab the store brand that tastes almost the same. You still eat breakfast. You just don't trigger the coupon police.

Automation's dirty secret: it can only control the accounts it can see

Tax-loss harvesting is often marketed as an automated feature, and it is. But it has a fragile dependency: *your other accounts.*

The wash sale rule doesn't care whether the repurchase happened in the same brokerage account. If you sell Fund A for a loss in your robo-advised taxable account and then your spouse buys Fund A in their separate brokerage account two days later, you may have created a wash sale problem for yourself.

Even more annoyingly, a purchase can happen without you noticing. Dividend reinvestment plans (DRIPs) can automatically buy more shares. A recurring contribution can auto-buy the same fund. Your 401(k) might not buy the exact same ETF, but if you're doing a lot of similar investing across accounts, you can stumble into "substantially identical" territory in ways that feel unfairly technical.

The platform can avoid wash sales *inside* the accounts it manages. It cannot stop you from unknowingly stepping on the rake in another account it can't see.

This is one reason some investors get less benefit from TLH than the marketing suggests. Their financial life is fragmented. The algorithm is only piloting one plane while you're simultaneously flying three other planes with autopilot settings you forgot you turned on.

Where the ``tax alpha" comes from (and what it is not)

People sometimes call the potential benefit of TLH "tax alpha," borrowing the investing term *alpha* (returns above a benchmark). It's a seductive phrase because it hints at outperformance without extra risk.

But TLH is not free money. It's better described as a timing shift and a rate arbitrage that *may* work in your favor:

- *Timing shift*: You recognize losses sooner, which can reduce taxes now. You might recognize gains later.
- *Rate arbitrage*: In some situations, you can end up offsetting gains taxed at one rate with losses while later realizing gains taxed at another rate, depending on holding periods and future tax circumstances.

The benefit depends on your current and future tax rates, your ability to use the losses, and whether the strategy accidentally creates unnecessary gains later. It also depends on whether you stay invested and allow the market to recover. Harvesting losses and then panic-selling your substitute is like collecting a coupon and then throwing the groceries away.

A clear way to think about it: TLH doesn't change what the market returns. It changes what *you keep* after taxes, sometimes by a little and sometimes by more, depending on your situation.

And, crucially, it works best when you have volatility. Smooth upward markets don't create many losses to harvest. TLH is a strategy that feeds on short-term misery and pays off in long-term patience.

The reason humans don't do this manually: it's tedious and it feels wrong

Even investors who understand TLH often don't implement it, because it combines two kinds of unpleasantness:

- *Administrative pain*: tracking lots, avoiding wash sales, picking substitutes, monitoring across time.
- *Emotional pain*: selling something right after it has disappointed you.

The administrative pain is solvable with software. The emotional pain is also solvable with software, in a way that sounds almost too simple: software does not feel shame. It does not interpret a sale at a loss as a verdict on your intelligence. It interprets it as a chance to improve after-tax results while maintaining exposure.

In that sense, TLH is like automated rebalancing with a different motive. Rebalancing fights drift. TLH fights taxes. Both are forms of disciplined rule-following in a market that constantly tries to provoke impulsive behavior.

A realistic example: the ``same risk, better tax outcome" story

Suppose you invest $50,000 into a diversified U.S. stock ETF in a taxable account. A few months later, the market dips and your position is down $5,000.

A platform with TLH might sell that ETF, realize the $5,000 loss, and buy a different U.S. stock ETF that provides similar broad-market exposure.

From your perspective, your portfolio doesn't suddenly become "cash." You remain invested in U.S. stocks. If the market recovers, you recover.

The difference is in the tax ledger: you now have a realized capital loss that can potentially offset other realized gains—maybe from selling a different investment, maybe from rebalancing, maybe from some other financial event. If you don't have gains this year, you may carry the loss forward under tax rules and use it later.

Notice what's missing: there is no claim that TLH prevents the loss. You still lost money in the market. TLH just tries to make the loss do one more job before it leaves the building.

The trade-offs and the gotchas platforms don't headline

TLH is powerful enough that it deserves a sober look at its friction points:

You can run out of losses. In a long bull market, the opportunity set shrinks. If everything you own is up, there's nothing to harvest. TLH is episodic, not constant.

You can accidentally create gains later. Harvesting a loss often means buying a substitute at a lower price. When the market recovers, those replacement shares may carry larger future gains. TLH can shift taxes into the future, not eliminate them. That can still be valuable, but it's not magic.

Short-term versus long-term matters. Holding periods affect tax rates on gains. TLH interacts with this in ways that can be beneficial or neutral depending on the timing and your broader tax picture.

Too much trading can create other costs. Even with commission-free trading, there can be bid-ask spreads and market impact (usually small for large, liquid ETFs, but not always). A well-designed TLH system tries to avoid harvesting tiny losses that aren't worth the friction.

It's not the same as tax advice. TLH is a mechanical strategy. It doesn't automatically mean the platform has modeled your entire tax return, your state taxes, your AMT considerations, or the way your income might change next year. Treat it as an optimization tool, not a holistic tax plan.

The best mindset is: TLH can improve after-tax outcomes at the margin, especially for investors in taxable accounts who have enough assets and volatility for it to matter. It is not a substitute for choosing the right account types, funding retirement accounts appropriately, and keeping a coherent overall plan.

How to tell whether TLH is likely to matter for you

There's a simple, slightly blunt question that cuts through the hype:

Do you have a meaningful taxable investing balance, and do you pay meaningful taxes?

If most of your investing is inside tax-advantaged accounts (like 401(k)s and IRAs), TLH is largely irrelevant there because the tax mechanics that make harvesting useful don't apply in the same way.

If you do invest in taxable accounts, TLH tends to be more relevant when:

- you are in a higher tax bracket,
- you have regular gains (from rebalancing, selling assets, or other taxable events),
- you have enough account size that harvested losses can be meaningful,
- you can avoid wash sales across your broader financial life.

The last bullet is the one most people underestimate. The easiest way to sabotage TLH is to run multiple overlapping strategies across multiple accounts without coordinating them.

Actionable takeaway: if you use TLH, turn off the ``silent wash sale generators"

If you want automated loss harvesting to work the way it's supposed to, do this audit once:

1. In every account you control (including accounts at other brokerages), look for dividend reinvestment settings on holdings that overlap with your TLH account. Consider turning DRIP off for those overlapping funds so you don't accidentally repurchase during the wash sale window.
2. Review any recurring purchases (automatic deposits, scheduled buys) that might buy the same or substantially identical fund you're harvesting. Pause or redirect them if needed.
3. If you share finances with a spouse or partner, agree on one sentence that prevents accidental sabotage: *"If we*

buy broad index funds in one place, we won't buy the exact same fund in another account without checking."

The point isn't to micromanage your life. It's to stop the background autopilots from undoing the one autopilot you intentionally turned on. TLH is a clever machine, but it's still operating inside the boundaries of what it can see—and what you don't accidentally break.

Chapter 5

The Market Machine and Your Mind

You tap 'Buy' on your phone, and confetti explodes on the screen. It feels like a game, but the money is real. In the split second between your tap and the trade confirmation, your order travels through a complex, high-speed labyrinth where milliseconds matter and invisible tolls are paid. But the biggest adversary isn't the machine; it's your own biology. This chapter explores the dangerous intersection where sophisticated market algorithms meet ancient human instincts, revealing how the financial industry often hacks your psychology to profit from your impulses.

5.1 What Actually Happens When You Click Buy?

You tap *Buy*. The screen does its little shimmy, maybe flashes a reassuring checkmark, and—emotionally—your brain files the whole thing under *done*. In the physical world, buying something usually means an exchange: you hand over money, you walk away with a thing. Markets are not a farmers' market. That tap is more like slipping a note into a machine room full of conveyor belts, trapdoors, and referees. The impressive part isn't that it works. The impressive part is that it works *fast*, for millions of people, all day, with prices that update in the blink between your intention and your confirmation.

To make this make sense, forget the romantic idea that stocks trade because two humans agree on a price at the same moment. Most of the time, what you're really buying is *liquidity*: the ability to convert your decision into an executed trade right now. Liquidity is the market's promise that you can get in (or out) without a long awkward wait. And like every promise worth having, it comes with terms.

The first thing you're buying is time

Imagine you want to buy 10 shares of a big, heavily traded company—something boring and gigantic. You'd think that's the easiest transaction in the world. But even in that "easy" case, the market has a problem to solve: who's on the other side *at this instant*, at a price you'll accept?

If you placed your order and had to wait until another investor happened to wake up, open their account, and decide to sell you exactly 10 shares at exactly your desired price, trading would feel like ordering a custom sofa. Instead, most of the time, you get something closer to ordering bottled water: it's already sitting on a shelf, priced, ready to go.

That "shelf" is called *the order book* when it's on an exchange, and a *quote* when it's provided by a market maker. Either way, the concept is the same: prices are posted in advance so you don't have to negotiate in real time.

Bid, ask, and the tiny toll booth you don't notice

Here's the basic language of that shelf:

- The *bid* is the best price someone is currently willing to pay to buy.
- The *ask* (or *offer*) is the best price someone is currently willing to accept to sell.

The difference between them is the *bid-ask spread*. If the best bid is $100.00 and the best ask is $100.02, the spread is two cents.

Two cents sounds like nothing until you realize what it represents: the market's toll for immediacy. If you want to buy *right now*, you typically pay the ask. If you want to sell *right now*, you typically accept the bid. The spread is the little gap between "I want it now" and "I'm willing to wait."

A useful way to think about the spread is as the price tag on certainty. When markets are calm and a stock is highly liquid, that certainty is cheap. When markets are stressed, uncertain, or thinly traded, it gets expensive in a hurry—spreads widen, and the toll booth starts collecting real money.

Your order: a tiny instruction with big consequences

When you click buy, you're not just saying "I want this stock." You're giving an instruction with optional fine print. The most common instructions look like this:

- *Market order*: "Buy as soon as possible at the best available price."
- *Limit order*: "Buy, but only if you can get $X or better."

A market order prioritizes speed. A limit order prioritizes price. Both are reasonable. Both can surprise you.

A market order can fill at a price slightly different from what you saw on screen a moment ago, especially in fast-moving conditions. That's not always "the broker cheating you." Often it's the simple fact that the price you saw was a snapshot, and the market is a flipbook.

A limit order sounds safer—and often is—but it carries a different risk: not getting filled. You can be *right* about value and still *not trade* because the market never touches your limit. Or it touches it briefly, fills some shares, and leaves you half-executed, like being able to buy only half your groceries before the store abruptly closes.

The deeper truth is that "the market price" is not a single number floating in the air. It's a constantly updated negotiation between buyers and sellers, and your order is one more move in that negotiation.

Where does your order go? The market is not one place

Most people picture "the stock market" as a building. In reality, it's a network: multiple exchanges, multiple trading venues, and multiple firms standing ready to trade. Your broker sits at the edge of that network like an air-traffic controller, and your order is a plane that needs a runway *now*.

Your broker has choices. It can route your order to an exchange, where it may interact with the visible order book. Or it can route the order to a market maker (sometimes called a *wholesaler*), a firm that internalizes many retail orders by filling them out of its own inventory or by matching them against other flow.

This is where the story gets interesting, because "who fills your order" is not just trivia. It can influence:

- how fast you get a fill,
- whether you get a tiny bit of price improvement,
- and how much the trade really costs you when you include the spread and any slippage.

And yes, it also raises incentives questions. Retail order flow is valuable. It tends to be smaller, less informed, and more predictable than professional trading. In plain English: it's often easier for sophisticated firms to trade against without getting ambushed.

Meet the market maker: the business of standing ready

A market maker is the opposite of a hesitant shopper. Its job is to be willing to buy and sell—continuously—so that other people can trade when they want. It provides liquidity the way a taxi stand provides rides: you show up, you don't have to find a driver, one is already waiting.

But taxis don't idle for free. The market maker gets paid primarily through the spread (and through other mechanisms that can exist in modern routing economics). If it can buy at the bid and sell at the ask often enough, and manage risk well enough, it can earn a living by being the "always open" counterparty.

This sounds harmless until you notice the hidden difficulty: if a market maker is always ready to trade, it is always at risk of trading with someone who knows something it doesn't. Maybe a big institution is unloading shares because of breaking news. Maybe a wave of orders is about to move the price. The market maker's fear is simple: *I sell at $100.02, and one millisecond later the real price is $99.70.*

That fear is why spreads exist. The spread is not only a fee for service. It's also insurance against being picked off by faster or better-informed traders.

High-speed algorithms: not evil, not magical---just relentless

Now add the modern twist: most market making isn't done by a person deciding, "Hmm, I'll buy 500 shares here." It's done by algorithms that:

- monitor many venues at once,
- update quotes constantly,
- manage inventory risk (how many shares they're holding),
- and react to incoming orders in microseconds.

If that sounds like science fiction, here's the human translation: these firms are trying to keep their prices competitive while avoiding being the last one to realize the world changed.

Picture a busy sandwich shop that must update its menu price every time the cost of bread changes—and bread costs can change 100 times a second. That's what a market maker is doing with risk. The "ingredients" are information and order flow, and the "menu" is the bid and ask it's willing to show the world.

A lot of popular commentary frames high-speed trading as a predator circling retail investors. The reality is more subtle. High-speed algorithms are everywhere because the market's plumbing demands speed. Quotes appear and disappear quickly. Opportunities for arbitrage—tiny price differences across venues—get closed quickly. And the competition to provide liquidity at a slightly better price is intense precisely because those pennies add up.

The uncomfortable part is that, while the system is competitive, it's also asymmetric: the firms designing microsecond strategies are not doing this on a phone over coffee. They're doing it with specialized infrastructure, teams of engineers, and an obsession with latency—the time it takes for information to travel.

A simple timeline of your ``instant" trade

A typical retail order often looks like this, in human time:

1. You place an order in an app.
2. Your broker receives it, checks basic constraints (account status, buying power, order size limits, risk controls).
3. The broker's routing logic decides where to send it (exchange or market maker).
4. The order is executed: you receive a fill report-price and quantity.

5. Later, behind the scenes, the trade is cleared and settled through industry infrastructure.

Notice the psychological trick: step 4 feels like the finish line, because it's when your app updates and your brain relaxes. But operationally, step 5 is where the system makes good on the promise that you really bought something.

Execution is the moment of agreement. Settlement is the moment of *delivery*. Those are different events, and the difference matters more than people think—especially during volatile markets, when plumbing stress can become headline news.

Why the price you see isn't always the price you get

You've probably experienced this: you watch a quote, you tap buy, and the fill comes back a cent or two higher. Or you try to sell, and you get a little less than expected. Sometimes you'll see the opposite: a fill slightly *better* than the displayed price, called *price improvement*.

This isn't necessarily manipulation. It's mostly the product of three very ordinary facts:

- *Prices move*: even tiny shifts matter when you're demanding immediacy.
- *Quotes are fragmented*: different venues show different liquidity at different moments.
- *Your order changes the world*: even a small order can consume the best available price if there are limited shares posted there.

The phrase *slippage* is used for the gap between the price you expected and the price you got. Slippage can be a couple of pennies, or it can be dollars in illiquid names or fast markets. It's not a moral failure. It's the market charging you for moving from "thinking about trading" to "trading."

Liquidity has a personality: deep, shallow, and suddenly gone

Liquidity isn't a static property like "this stock is liquid." It's a living condition. At 10:30 a.m. on a quiet Tuesday, a large stock may have thick liquidity—many shares posted at many price levels. Your market order barely makes a ripple.

At 9:30 a.m. right after the opening bell, or during a surprise earnings announcement, liquidity can thin out dramatically. That's when spreads widen and price gaps appear. It's not that the market stops working. It's that the market becomes honest about uncertainty.

A vivid way to picture this is to imagine walking across a river on stepping stones. In calm conditions, stones are close together, and you don't think about it. In turbulent conditions, stones are farther apart. You can still cross—but now each step costs attention, and a rushed step can get you wet.

That's what happens to execution quality when volatility spikes. The price you get becomes more sensitive to timing, order type, and routing.

Why tiny retail orders are treated differently than big institutional ones

A retail trade for 10 shares looks small to you because it is small to you. To the market's machinery, small orders have a different statistical profile than a pension fund buying 200,000 shares.

Institutions often split large orders into smaller pieces to avoid moving the price against themselves. They may use algorithms designed to hide intent and minimize market impact. Their problem is: *if the market senses I'm a huge buyer, it will raise prices before I finish.*

Retail orders, on average, are less about stealth and more about immediacy. They arrive as a stream of small decisions: a buy here, a sell there. Because that flow is comparatively predictable, firms compete aggressively to interact with it.

This is one reason modern market structure looks like it does: retail flow has become its own ecosystem, with specialized firms and routing logic designed to capture and monetize it—sometimes by giving you a slightly better price than the displayed quote, sometimes by simply being the fastest and most reliable counterparty.

The spread is not the only cost, but it's the most honest one

If you want a single number that represents the "cost" of trading, the spread is the cleanest starting point because it's mechanically tied to how liquidity works. You can't wish it away with marketing. You can't argue it into zero. The market either offers you an immediate counterpart at a tight spread, or it doesn't.

That said, what you *pay* isn't always equal to the posted spread. The cost you actually experience is often described as an *effective spread*: a measure of how far your execution price was from a fair reference point (often the midpoint between bid and ask at the time). You don't need to calculate effective spreads by hand to benefit from the concept. You just need the mindset shift:

> Every trade has a price, even when the commission is $0.

The bill doesn't always show up as a line item. Sometimes it shows up as a slightly worse fill. Sometimes it shows up as the trade you *didn't* get because you used a market order in a moment when liquidity briefly vanished. Sometimes it shows up as "I traded a lot this year and can't quite explain why my results feel a little...sticky."

A practical example: buying calm vs. buying chaos

Take the same order—buy 50 shares—and run it through two different moments.

Calm market, liquid stock. The spread is a penny or two. Your order is filled almost instantly, and if you look closely you might even get fractional price improvement. The execution feels like what the app promised.

Chaotic market, thin liquidity. The spread widens. Quotes flicker. Some liquidity pulls back because no one wants to be the first to quote a "wrong" price. Your market order becomes a battering ram: it takes whatever is available, and "available" may be worse than the last price you saw. The execution feels like betrayal, even though the market is simply charging the going rate for certainty.

Same button. Same intention. Completely different experience. That difference is the market's microstructure—the small, mechanical details that determine whether trading feels like frictionless shopping or like trying to buy concert tickets during a stampede.

So what should a normal person do with this information?

You don't need to become a part-time market-structure nerd to protect yourself. You just need to stop treating the *Buy* button as a magical portal and start treating it as a request to the market's liquidity engine.

Here's the actionable takeaway that pays for itself the fastest:

> *When you care about price, use a limit order—and when the market is moving fast, slow yourself down on purpose.*

A limit order forces you to name your price. It turns "whatever" into "only if." That single constraint can protect you from the worst moments of slippage, especially in volatile stocks, at the open and close, or during major news. If your limit doesn't fill, that's not failure—it's information. It's the market saying, "The instant version of this trade costs more than you're willing to pay."

5.2 The Price of `Free': Hidden Costs and Compounding

"Commission-free" is one of those phrases that lands in the brain like a warm blanket. No fee. No friction. No penalty for changing your mind. The button is right there; the market is right there; why *not* take a little swing?

Because markets have a sense of humor: when a product advertises that it costs nothing, it usually means the price moved somewhere you're less likely to look. Not necessarily in a villainous way. Often in a clever, engineered way—small enough per trade that you can ignore it, consistent enough over time that it can quietly dominate your results.

The trick is that most investing costs don't arrive as an invoice. They arrive as *math*. A slightly worse price. A slightly wider spread. A little more cash sitting idle. A little more trading because it feels painless. Each one is easy to shrug off. Together, they can behave like gravity.

``Free" changed what you notice---and what you don't

The old brokerage model was blunt: you paid a commission, you winced, you traded less. The pain was visible. It felt like paying for checked baggage at the airport: annoying, but at least you knew when it happened.

Zero-commission trading did something more psychological than financial: it removed the immediate sting that used to act as a speed limit. When the cost of pushing the button looks like $0.00, your attention stops scanning for costs. You start scanning for *opportunities*. The app becomes a slot machine that happens to show candlestick charts.

Meanwhile the costs that remain—the ones that never needed your permission—are still there, doing their work in the background. The market still charges for immediacy. Dealers still get paid to provide liquidity. Brokers still have to run a business. And the incentives didn't disappear. They just became more like plumbing: hidden behind the wall, extremely consequential when you understand how it's arranged.

The spread: the market's ``service fee" that isn't called a fee

Every time you trade, you run into the bid-ask spread: the difference between the best price buyers are offering (the bid) and the best price sellers are asking (the ask). The spread is the price of liquidity—what you pay to do something *now* instead of waiting.

Even if your broker charges $0 commission, the spread is still there. In a liquid stock it might be a penny. In a thinly traded stock it might be a dime, a quarter, or a full dollar. In a market panic it can widen like a mouth opening to swallow uncertainty.

Here's the uncomfortable part: the spread is not just a market fact. It's a profit opportunity for the firms that consistently sit between buyers and sellers. Market makers compete to capture it. Exchanges compete to attract orders. Brokers route orders through a maze of venues. And when people say "the app is free," what they really mean is: *the app is free to you because someone else is monetizing your order in the machinery that determines your execution price.*

That doesn't mean you're always getting a bad deal. It means you're participating in a deal you rarely see.

Payment for Order Flow (PFOF): selling the right to trade with you

Payment for Order Flow—usually shortened to *PFOF*—is the arrangement where a broker receives compensation for routing customer orders to a particular trading firm (often a wholesaler/market maker) rather than sending them directly to an exchange.

Translated into normal language: your broker can get paid for sending your order to someone else to fill.

This makes people understandably uneasy. It sounds like the referee taking tips from one team.

The defense is also understandable: wholesalers often fill retail orders quickly and can sometimes offer tiny price improvements versus the publicly displayed quote. They argue they're competing for your trade by giving you a slightly better price than what you would have gotten elsewhere, and the broker passes on "free trading" to you.

Both things can be true at once:

- PFOF can help subsidize low explicit costs.
- PFOF creates a conflict of interest because routing decisions can be influenced by who pays the broker, not just by what gets you the best outcome.

That word—*conflict*—matters. In a well-run system, conflicts are managed with rules, disclosures, and oversight. In a poorly understood system, conflicts become a design space. You don't need to assume bad intent to be cautious. You just need to accept that incentives shape behavior the way gravity shapes water.

Best execution: the promise that ``fast" isn't the same as ``best"

Brokers are generally obligated to seek *best execution*: to use reasonable diligence to get you the most favorable terms under the circumstances. That's more than a slogan. It's an idea with teeth: price, speed, likelihood of execution, and the overall quality of the fill all matter.

But best execution is not the same as "you got the National Best Bid and Offer." Getting an execution at the publicly displayed best price can still leave room for questions:

- Could you have received more price improvement on another venue?
- Did the order get filled quickly because it was routed to a venue that pays the broker more?
- Did the routing decision optimize for the broker's revenue, or for your outcome?

The core problem isn't that any one trade is outrageously expensive. It's that the difference between "pretty good" and "slightly better" can become *enormous* when repeated across years of contributions, rebalancing, and life events.

In other words: execution quality is a rounding error that gets promoted to a major character through repetition.

The cost you feel least is the cost that compounds the most

People obsess over the big, dramatic costs: the market crash, the one terrible stock pick, the tax surprise. Those matter. But the most reliable wealth destroyers tend to be boring. They're the frictions that happen so regularly you stop noticing them.

A one-time 20% loss is painful—but it's also a moment you respond to. You rebalance. You change habits. You talk about it.

A 1% annual drag is quieter. It doesn't punch you. It whispers. It doesn't make headlines. It just skims a little off the top every year—forever. And because the market's gift to patient investors is compounding, a small annual subtraction doesn't merely subtract. It changes the slope of your future.

That is the real price of "free": it can encourage you to ignore the low-decibel costs that never stop running.

Compounding friction: why 1% is not ``basically nothing"

Compounding is usually introduced as a friendly miracle: money earns returns, returns earn returns, and time does the heavy lifting. That's true. It's also incomplete.

Compounding is a machine that amplifies whatever you feed it—good or bad. If you feed it a little extra cost every year, it doesn't merely collect that cost. It compounds the *absence* of what you could have had.

Here's the cleanest way to see it. Suppose you invest $10,000 for 30 years and earn 7% annually. Now suppose everything is the same, except there's a 1% annual drag from some combination of costs: wider effective spreads, cash sitting uninvested, slightly worse execution, higher turnover nudged by "free" trading, or all of the above. Your net return is 6%.

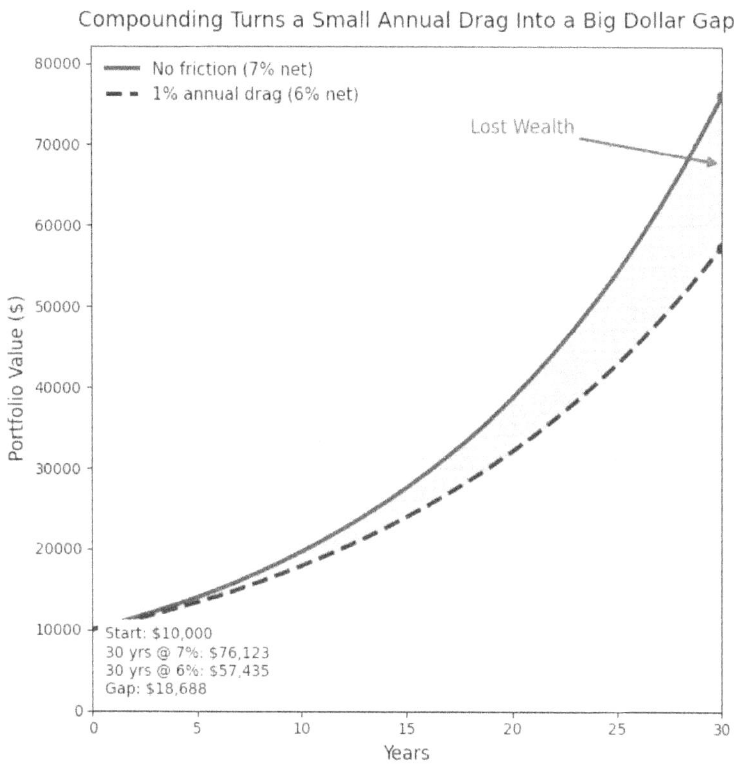

Look at the gap, then look at what caused it: not a crash, not a scandal, not a once-in-a-decade catastrophe. Just a persistent 1% drag—something many investors could easily "afford" in their heads because it doesn't feel like spending.

And notice what's most infuriating about this kind of loss: it's not dramatic enough to trigger your defenses. There's no emotional alarm. It's just... less. Less growth. Less flexibility.

Less margin for error. Less ability to take a career break, help a family member, move, retire, or sleep.

Where the drag comes from (even when the commission is $0)

The phrase "hidden costs" can make it sound like there's one secret fee. In reality it's a handful of small frictions that can stack:

Execution and spreads. You pay the spread when you cross from bid to ask. If your execution is consistently a hair worse than it could have been, you'll never see a bill—but you'll feel it in long-run performance.

Turnover you didn't need. When trading feels free, you treat ideas like disposable cups. Buy. Sell. Rotate. Chase. Even if each trade has only a tiny cost, more trades means more tiny costs. The market doesn't need to beat you. It just needs you to keep volunteering tolls.

Cash drag. Many brokerages sweep uninvested cash into vehicles that pay you something-sometimes very little-while the broker earns more on the spread between what they make and what you receive. Cash is part of a portfolio, whether you pay attention to it or not. A low yield on idle cash is an annual drag that feels like "nothing happened," which is exactly why it's so effective.

Behavioral drag. Even if your broker executed every trade perfectly, your own behavior can generate a cost that looks eerily similar to a fee: buying when excitement is high, selling when fear is high, trading because the app nudged you at lunch.

The zero-commission era didn't invent that problem, but it removed a natural brake.

You don't have to prove which friction is biggest to take the right lesson. The lesson is simply that *friction exists,* and the investor who plans for it has an advantage over the investor who only plans for market returns.

A subtle point: ``price improvement" can be real and still not settle the conflict

You'll sometimes hear that wholesalers provide price improvement-executions at a slightly better price than the displayed best bid or offer. That can be true, and for many retail-sized orders it often is. But "price improvement exists" doesn't end the conversation, because the questions that matter to you are comparative:

- *How much* improvement did you get on average?
- Would you have received *more* improvement on a different venue?
- Are you getting good outcomes *consistently,* or only when it's easy?

It's possible for an execution to look good on paper and still be part of an ecosystem that encourages overtrading. It's possible for the average fill to be fine while the worst fills happen at exactly the moments you're most likely to panic. "Free" doesn't have to be a scam to be expensive.

How the ``free" model changes product design

A brokerage that earns revenue primarily from trading-related mechanisms has a completely different appetite than a brokerage that earns revenue mainly from a transparent advisory fee.

When trading drives revenue, the business quietly prefers that you:

- trade more often,
- keep more cash on the platform,
- and treat investing as an activity rather than a plan.

That preference doesn't require anyone to do anything unethical. It can show up as product design:

- defaulting you into market orders,
- highlighting "top movers,"
- making options trading feel like a feature unlock,
- making the app feel alive when you're active and silent when you're patient.

If you want to understand how costs leak into your life, look for places where the product celebrates activity for its own sake. The market already provides enough stimulation. You don't need your broker adding caffeine.

What to do with this: a practical ``cost control" checklist

You can't negotiate with the existence of spreads. You can't personally redesign market structure. You can, however, make a few decisions that dramatically reduce the odds that "free" becomes expensive.

- *Trade less, but with more intent.* If you're investing for long-term goals, frequent trading is usually a cost center, not a skill signal.
- *Use limit orders when spreads are wide or volatility is high.* A limit order is you telling the market, "I'm not paying panic prices today."
- *Pay attention to cash yields and sweeps.* Uninvested cash is part of your portfolio. If it's earning close to nothing while rates are meaningfully higher elsewhere, that difference is an annual fee in disguise.
- *Read your broker's order-routing and execution-quality disclosures-once.* You're not looking for perfection; you're looking for whether the firm is transparent and whether the incentives are clearly explained.

None of this requires paranoia. It's the same mindset you'd apply to any "free" service that touches something important. You wouldn't let a "free" contractor remodel your kitchen without understanding how they're paid.

Actionable takeaway

Pick one account you actively use and do a 15-minute "friction audit" today:

- Turn on a default that makes you choose *limit* vs. *market* each time (or set limit as the default if your platform allows it).

- Check what your idle cash is earning; if it's materially lower than widely available alternatives, decide whether you'll keep less cash there.
- Look at your last three months of trades and ask one blunt question: *If each trade had a $5 commission, would I still have made it?* If the answer is no, "free" is already costing you.

The goal isn't to find a perfect broker or a perfect market. The goal is simpler: stop paying invisible, compounding costs for entertainment—and reserve your money's compounding power for you.

5.3 Why Your Brain is Bad at Trading

The most dangerous thing in your brokerage account isn't options, leverage, or some shadowy algorithm with a fiber-optic backbone. It's the part of you that wants to *feel* like a competent person. Markets are brutal not because they're complicated, but because they keep score in money while your brain keeps score in emotions.

You can know, intellectually, that the price you see is a moving negotiation and that every trade has friction. Then your stock drops 4% before lunch and your body reacts as if you've heard footsteps behind you on a dark street. Heart rate up. Jaw tight. A quick scroll for reassurance. A hand hovering over *Sell* like it's a fire alarm. That response isn't a personal flaw. It's ancient software.

Humans didn't evolve to compound wealth. We evolved to avoid becoming lunch.

Your brain's job description: survive today, not retire in 30 years

The investing horizon that matters most-decades-barely registers on a nervous system built for afternoons. Your ancestors who treated rustling grass as "probably nothing" didn't get to pass on their calm temperament. The anxious ones lived long enough to have kids.

Markets exploit that wiring because markets are a machine that produces ambiguity on purpose. Prices move even when nothing "happens." News arrives in fragments. Other people act for reasons you can't see. And the scoreboard updates constantly. That's basically a perfect environment for cognitive shortcuts-mental rules of thumb that are great for survival and terrible for trading.

A useful mental reset is to stop asking, "Why do I do this?" and start asking, "What was this impulse *for* in the Stone Age?" The answer is almost never "to maximize risk-adjusted returns."

Loss aversion: why a small dip can feel like an emergency

Loss aversion is the simplest, most expensive bias in investing: losses hurt more than gains feel good. Not "a little more." Often *a lot* more. Your account being up $500 can feel mildly pleasant. Being down $500 can feel like an insult, a threat, and a referendum on your intelligence.

This creates a trap that looks like prudence from the inside. When prices fall, your brain doesn't think, "expected volatility." It thinks, "damage." That word brings urgency, and urgency creates action. In markets, action is often the costly choice.

Loss aversion also twists the way you remember. After a loss, you don't just remember the outcome-you remember the sensation of being wrong. Your brain then starts seeking an antidote: get the money back, get the feeling back, restore the story where you're competent and in control. That's how perfectly normal investors end up making emotionally logical decisions that are financially irrational.

It's not that you can't handle risk. It's that your brain interprets risk as an immediate physical problem.

The disposition effect: the weird habit of selling winners and ``giving losers time"

If loss aversion had a signature move, it would be this: selling investments that went up (to "lock in" a win) while hanging on to investments that went down (to avoid "making it real"). Researchers call it the *disposition effect*. Regular people call it "being sensible," because it feels responsible to take profits and patient to wait for a loser to recover.

The market doesn't care what feels responsible. The market cares what happens next.

Selling winners early often means you cut off the right tail-the part of investing returns that comes from a small number of big winners over time. Holding losers too long often means you keep money trapped in the emotional project of *not being wrong*. Sometimes a loser rebounds. Sometimes it doesn't. The disposition effect isn't about predicting which will happen. It's about how your brain quietly changes the rules based on your ego.

A particularly nasty version shows up when people "average down" on a loser not because the fundamentals improved, but because the lower price feels like a bargain compared to their purchase price. That purchase price becomes a psychological anchor, and the investment becomes a story about *getting back to even*. The market is happy to sell you that story in installments.

Anchoring: your purchase price becomes an emotional home address

Anchoring is what happens when an irrelevant number becomes a reference point your brain won't stop using. In investing, the most common anchor is your entry price: "I bought at $50." From that point on, your brain treats $50 like a fair price in a moral sense, not just a historical fact.

If the stock is $45, it feels "cheap." If it's $55, it feels "expensive." That feeling has almost nothing to do with the company's value and almost everything to do with the human tendency to treat a first number as the center of the universe.

Anchoring makes charts seductive. A chart is basically a factory for anchors: highs, lows, round numbers, yesterday's close. You can stare at them until you feel informed while learning very little about what actually matters. The longer you look, the more you feel that the stock is "supposed to" return to some familiar level-as if prices have a homing instinct.

Prices don't have homes. They have auctions.

Recency bias: why the last month tries to impersonate the future

Recency bias is the brain's shortcut of assuming that what just happened will keep happening. It's efficient in everyday life. If the stove burned you yesterday, you treat it as dangerous today. Markets punish that efficiency.

When markets rise for a while, recency bias turns optimism into a baseline. A steady upward chart starts to feel like a law of nature. "Every dip gets bought." "This time is different." When markets fall, the bias flips into dread. A red week becomes a red year in your imagination. You start seeing risk everywhere, like a person who once hit black ice and now drives with white knuckles all winter.

Recency bias is why people buy after good performance and sell after bad performance-essentially reverse-engineering "buy high, sell low" while feeling like they're being responsive.

It's also why investors tend to talk about the market as if it has moods. Really, it's our minds that have moods, and we project them onto price movements like a movie onto a screen.

Overconfidence: the quiet belief that you're the exception

Overconfidence in investing rarely looks like someone shouting, "I'm a genius!" It usually looks like something gentler: "I've done my homework." "I understand this company." "I can tell when something is overhyped." "I'll know when to get out."

Overconfidence is dangerous because it doesn't need to be extreme. A small surplus of confidence over skill is enough to generate extra trading. Extra trading creates extra friction. And friction is one of the few costs that will happily accept as many deposits as you want to make.

There's strong empirical evidence that the most active individual traders tend to underperform less active investors-often substantially-after costs. The details vary by dataset and era, but the direction is stubborn: trading more tends to mean earning less. The market is not impressed by your effort. The market charges by the transaction.

A particularly sharp edge of overconfidence is believing you can time your emotions. People think they'll panic-sell only if things are "really bad." Then "really bad" arrives dressed as a normal pullback plus scary headlines, and the body takes over. You don't time fear. Fear times you.

Confirmation bias: how your feed becomes your financial advisor

Confirmation bias is the tendency to seek information that agrees with what you already believe and dismiss information that challenges it. In investing, it's gasoline.

Once you own something, you become motivated to believe it was a good decision. Your brain starts curating evidence like a defense attorney. You find the bullish analyst. You save the optimistic chart. You join the subreddit where everyone speaks your hope out loud. You interpret neutral news as positive because you want it to be.

The modern internet turns confirmation bias into a subscription service. Algorithms feed you what keeps you engaged, and nothing keeps you engaged like being told you were right. You begin to confuse repetition with truth. You start to treat agreement as research.

The cruel part is that confirmation bias can make you feel calmer while making you more fragile. The more one-sided your information diet becomes, the more shocking reality feels when it arrives.

Herding: why other people's certainty is irresistible

Herding is not stupidity. It's a social survival mechanism. If everyone is running, you run. The cost of running unnecessarily is small compared to the cost of not running when there's real danger. Herding is why humans built civilizations. It's also why bubbles and panics are so easy to manufacture.

In markets, herding shows up as the magnetic pull of "most popular" lists, viral tickers, and narratives that come with a ready-made community. A stock becomes more than a claim on future cash flows; it becomes an identity and a team sport.

When you're in a herd, you outsource doubt. That feels amazing. Doubt is tiring. The herd offers you emotional leverage: you don't have to be right alone. But the market doesn't grade on group projects. You can be wrong with millions of friends and still lose money the same way.

Herding also has a timing problem. The emotional peak of a herd-the moment it feels safest, loudest, most inevitable-is often close to the moment where the trade becomes most crowded and the future returns become most disappointing.

That's not mystical. It's arithmetic. If everyone already bought, who's left to push the price higher?

Availability bias: why vivid stories beat boring statistics

Availability bias is your brain overweighting information that's vivid, recent, or emotionally charged. A dramatic story about someone turning $5,000 into $500,000 in a month sticks in memory like a catchy song. The boring truth that most people who try that blow up their accounts is harder to picture, so it feels less real.

This is why speculative frenzies spread faster than sensible plans. A retirement contribution is a calendar reminder. A meme stock win is a legend. One is realistic. The other is cinematic. Humans are built for cinema.

Availability bias also explains why people fixate on the few spectacular blowups in diversified investing-"What if the market crashes?"-and ignore the far more probable danger: failing to invest steadily, paying too much friction, and reacting emotionally at the worst times.

The market's biggest risk to most long-term investors is not that it will be volatile. It's that volatility will talk you into self-sabotage.

The ``do something" reflex: when activity masquerades as competence

When you feel uncertain, doing something feels smarter than doing nothing. Activity creates the sensation of control. That's useful in a kitchen fire. In a portfolio, it can be disastrous.

Markets create a specific kind of discomfort: you can't *solve* uncertainty. You can only price it, endure it, or insure against it through diversification and time. That's psychologically unsatisfying. Your brain wants a lever to pull.

The financial industry is happy to offer levers: trade alerts, analyst upgrades, breaking news banners, "opportunity" notifications. Some of it is information. Some of it is entertainment wearing a tie.

A brutal question to ask yourself before any trade is: *Am I improving my long-term plan, or am I treating discomfort as a problem that requires movement?* If it's the second one, you're not managing money-you're managing feelings with money.

A more realistic definition of ``skill" for most investors

For professional traders, skill might mean short-term forecasting, execution tactics, and risk-taking under pressure. For most people, "skill" looks different and far less glamorous:

- Building a portfolio you can hold through ugly markets.
- Keeping costs and friction low.
- Not turning headlines into portfolio decisions.
- Staying invested long enough for compounding to matter.

That kind of skill doesn't create great dinner-party stories. It creates a better life.

It's also where your brain pushes back the hardest, because the reward is delayed and the work is mostly emotional restraint. If you want to see how difficult that is, try doing nothing while your phone shows your portfolio bouncing around like a heartbeat monitor. The difficulty isn't ignorance. It's physiology.

A concrete way to disarm your biases: build rules that don't negotiate with your mood

The reliable way to beat a bias is not to argue with it in the moment. In the moment, the bias is you. You're not going to "logic" your way out of adrenaline.

What works better is building simple rules in advance-rules that turn future emotional decisions into mechanical ones. The point isn't to become robotic. The point is to stop letting your most anxious self have trading permissions.

A few rules that are boring enough to work:

- Decide in advance what your portfolio is for (retirement, house, flexibility) and match risk to that purpose.
- Rebalance on a schedule (quarterly or annually) instead of rebalancing based on vibes.
- When you feel an urge to make a "quick trade," wait 24 hours. If it's still compelling, use a limit order and size it small enough that you can sleep.

These rules don't require forecasting. They require self-knowledge: you're not designing a strategy for the market you *hope* you'll be in. You're designing one for the nervous system you actually have.

Actionable takeaway

Write down one sentence on a sticky note (or in the notes app you actually open) and make it your personal trading speed limit:

> *"If I'm feeling rushed, offended, euphoric, or scared, I'm not allowed to trade today."*

Then add one practical backstop: set a default for contributions and investing that happens automatically (paycheck, monthly transfer, or whatever fits your life), so your long-term plan keeps running even when your short-term brain is having a bad day. The goal isn't to eliminate emotion. The goal is to make sure emotion can't reach the *Buy* button without passing through a rule you wrote while you were calm.

5.4 Gamification: The Casino in Your Pocket

Your phone already knows how to make you do things you didn't plan to do. Open it to check the weather, and twenty minutes later you're arguing with a stranger, watching a video you didn't choose, and somehow shopping for a thing you didn't know existed. That's not a lack of willpower. It's design.

Now place a brokerage account inside the same attention economy. Same device. Same notification system. Same thumb muscle memory. Only now the "engagement" they're chasing isn't just your time—it's your trades. And because trading has real financial consequences, the cost of being nudged is no longer a little lost productivity. It's your future compounding quietly getting picked apart in exchange for a short-term hit of feeling involved.

The scary part isn't that trading apps are *fun*. Fun is fine. The scary part is that "fun" can be used as camouflage for incentives. When a platform makes money when you do something, it has a reason to make that something feel irresistible.

The investor fantasy: information, control, and a clean win

Most people don't start trading because they love microstructure or discount rates. They start because investing is one of the few arenas where *being smart* feels like it should translate directly into money. That's a powerful fantasy: read, decide, click, win.

Gamified design pours fuel on that fantasy. It turns investing into a series of crisp, satisfying micro-decisions-tap, swipe, confirm-like a strategy game where the rewards are immediate and the penalties can be blamed on "bad luck" or "manipulation" instead of what they often are: the normal costs of trading plus normal human bias.

When you combine a brain that hates uncertainty with an app that offers constant action, you don't get a calm long-term investor. You get a person trying to buy relief.

What ``gamification" really means in finance

In a pure sense, gamification is just using game-like elements-points, badges, streaks, confetti, celebratory animations-to make an activity more engaging. In fitness, that can be harmless or even helpful. In language learning, it's often genuinely useful. In investing, it's playing with matches in a room full of gasoline.

Why? Because the healthiest investing behaviors are emotionally underwhelming:

- Make steady contributions.
- Diversify.
- Rebalance occasionally.
- Ignore most noise.

Those behaviors are not "sticky." They don't produce a daily dopamine drip. They don't create suspense. They don't make a company look like it's growing fast.

Trading does.

So if you run a business whose revenue is boosted by transactions or engagement, the temptation is obvious: redesign the experience so that *trading* feels like progress and *patience* feels like inactivity. That's how you get an app that can claim it "democratized finance" while quietly training people to confuse motion with mastery.

Dopamine isn't pleasure. It's pursuit.

People talk about dopamine like it's the chemical of happiness. A better everyday description is that dopamine is the chemical of *wanting*. It's what makes you reach for the next thing: the next swipe, the next refresh, the next outcome.

Trading apps don't need you to be happy. They need you to be *activated*. They need your attention in a state of anticipation.

That's why variable rewards are so effective. A slot machine doesn't pay out every pull; it pays out unpredictably. That unpredictability is the hook. In trading, the market provides the randomness for free:

- sometimes you buy and it pops,
- sometimes you sell and it drops right after (making you feel brilliant),
- sometimes you "almost" timed the bottom,
- sometimes you get a lucky fill.

The app doesn't have to rig anything. It only has to frame those random outcomes as *your* skill—and make sure you're there for the next pull of the lever.

The interface is not neutral: it's a steering wheel

A brokerage interface can be designed like a cockpit or like an arcade.

A cockpit design makes you slow down. It forces you to confirm order type, shows the spread clearly, makes risk hard to ignore, and treats trading like a serious action with consequences.

An arcade design makes you speed up. It removes friction. It uses bright colors, movement, rankings, "top movers," and social proof to create urgency. It makes the environment feel alive when you're active and boring when you're not—which is a subtle way of telling you what kind of user the company prefers.

Even small interface choices matter:

- Is the default order type a market order (fast, potentially sloppy), or a limit order (intentional)?
- Is profit shown in big green numbers while risk is buried in a submenu?
- Are percentage changes displayed like a scoreboard, updated constantly, begging you to check again?

Nothing about those choices is inevitable. They're product decisions. Product decisions create behavior. Behavior creates returns.

Push notifications: outsourcing your impulse control

Push notifications are a genius invention if your goal is to interrupt someone at the exact moment they were trying to do something else. They hijack attention with the authority of "important message," even when the message is basically, "Hey. Remember this thing? You could do something about it. Right now."

In a trading app, notifications can be informational—an order filled, a dividend paid, a document posted. That's fine.

The problem is promotional or emotional notifications: price alerts that turn normal volatility into a personal summons. "Your stock is moving!" "Biggest movers today!" "Don't miss out!" These messages don't merely inform you. They recruit you into an emotional state where action feels urgent and inaction feels negligent.

Your brain already has recency bias and availability bias. Notifications are a delivery mechanism for those biases. They take what would have been a passing market wiggle and deliver it as a tiny crisis to your pocket.

If you've ever felt your stomach drop because of a phone buzz, you understand the issue: it's hard to be a long-term investor while carrying a device that can trigger short-term fear on demand.

Celebrations and ``micro-rewards": confetti is not cute when money is involved

A celebratory animation after a trade is a psychological statement: *you did a good thing*. It treats a trade like an accomplishment rather than a tool. That matters because many trades are not accomplishments. Many trades are unforced errors.

The classic example is confetti-an app literally throwing a party when you buy or sell. Some platforms have toned down or removed those features after criticism, which tells you something important: people noticed the steering.

The deeper issue isn't confetti itself. It's the moral framing. A platform that celebrates every trade is saying, "More is better." A platform that treats trading like using a power tool—useful, but not for fidgeting—sends the opposite message.

Your portfolio doesn't care whether you had fun. Your portfolio cares how often you paid the spread, how often you got shaken out by noise, and how much time you spent being "active" instead of being invested.

Leaderboards, ``most popular" lists, and the social pressure to participate

Humans are social learners. We copy what seems to work for other people. In healthy contexts, that's efficient. In markets, it's combustible.

"Most popular" lists and trending tickers are not just information. They're social proof: evidence that *other people* are doing something. Social proof does two things at once:

- It reduces the discomfort of uncertainty ("I'm not alone").
- It increases urgency ("I might miss it").

That combination is basically the recipe for herding—people clustering into the same trades at the same time, often pushing prices away from fundamentals. Once a trade becomes a social event, it becomes harder to leave without feeling like you're betraying a group or admitting you were fooled.

And then the cruel part: when the crowd finally turns, it doesn't turn gently. Herds don't tiptoe. They stampede.

The attention trap: focusing you on what moves, not what matters

Trading apps often emphasize what is most *salient*: the stocks with the biggest daily moves, the biggest percentage gains, the most dramatic news, the loudest community.

But what matters for long-term wealth is usually what is least salient: low costs, diversification, tax efficiency, and behavior you can stick with.

This mismatch is not accidental. The market's most important truths are boring. The app's business needs you to keep looking.

So the platform highlights the kind of information that creates feelings-excitement, fear, envy, urgency-because feelings are the energy source for repeated engagement. And repeated engagement is the precondition for excessive trading.

That doesn't mean "never look at your account." It means you should be suspicious of any interface that makes you feel like you're behind if you're not constantly watching.

``Education" that's really a funnel

A lot of platforms mix genuine education with a subtle funnel toward more complex, higher-turnover products. You start with stocks. Then fractional shares. Then margin. Then options. Then "advanced strategies." Each step is framed as leveling up.

But complexity is not maturity. Often it's just complexity.

True education in personal finance tends to make you calmer. It gives you a plan and reduces the need to act. Education that increases your sense of urgency, makes you feel FOMO, or makes you feel inadequate unless you trade "like a pro" is often not education. It's marketing with a textbook accent.

If a platform's learning path ends with you doing more transactions, that's not automatically bad. It's just worth noticing who benefits if you graduate.

Where the incentives snap into focus

Earlier, the costs of trading were described as small, repeated frictions: spreads, slippage, and the subtle drag of being too active. Now connect that to business models.

If a broker earns more when you trade more—directly or indirectly—then "engagement" and "revenue" become aligned. The platform can honestly say it wants you to be engaged. It can also honestly report rising revenue when you are.

That alignment is the core danger of gamification in finance. It means the same design patterns that maximize screen time in social media can be repurposed to maximize trading activity in investing.

And because those patterns work best when you're emotional—excited, scared, angry, validated—you end up with a system where your least financially disciplined moments are also your most profitable moments for someone else.

How to tell whether an app is trying to help you invest or trying to make you trade

You don't need a psychology degree. You just need a few simple tells.

A platform that wants you to invest tends to:

- emphasize goals (retirement date, emergency fund, down payment) over tickers,

- show long-term performance more prominently than daily swings,
- make diversification easy and speculation feel like an edge case,
- treat silence as success.

A platform that wants you to trade tends to:

- emphasize today's movers,
- send frequent prompts,
- celebrate activity,
- make risk feel like a spicy personality trait rather than a real constraint.

Neither style guarantees good outcomes. But one style is more likely to leave you with an account full of stories and not much compounding.

A personal boundary that changes everything: separate ``investing" from ``entertainment"

One of the most practical defenses is also the most boring: decide whether your investing app is allowed to be entertainment.

If you need entertainment, there are endless options that don't charge you the bid-ask spread. The market is an extraordinarily expensive place to go for dopamine.

If you genuinely enjoy finance and markets, you can keep that interest-just route it into places that don't trigger trades:

- read,
- paper trade,
- build watchlists,
- study a company for curiosity,
- talk about ideas without needing to act on them.

Your portfolio will not reward you for being entertained. It will reward you for being consistent.

Actionable takeaway

Do one thing that makes your future self richer and your present self slightly less "engaged":

> *Turn off every non-essential trading notification, and keep only confirmations (fills, deposits/withdrawals, security alerts).*

Then add a second rule that cuts the link between emotion and action: only allow yourself to place trades during a small, scheduled window-say, one day a week, or one hour on Fridays. The market will still move without you. That's the point. You're training your brain to stop treating every price wiggle as a request for a response.

If "free trading" is the candy, your job is to stop eating candy whenever someone waves a bag in your face. Your money doesn't need more excitement. It needs fewer interruptions.

Chapter 6

Algorithmic Gatekeepers

Remember the scene in old movies where a couple sits across from a stern bank manager, hoping for a loan? That desk is empty now. Today, your financial fate-whether you get that mortgage, that car loan, or affordable insurance-is decided in milliseconds by an invisible judge you'll never meet. This chapter pulls back the curtain on the 'algorithmic gatekeepers' that analyze your digital footprint to decide what you pay and if you're worthy of risk.

6.1 From Handshakes to Heatmaps: The AI Credit Upgrade

A generation ago, "getting credit" could look like a small-town ritual: you walked into a bank, sat across from someone who knew your last name, and left with a yes or a no that felt suspiciously like a judgment about your character. Even when credit cards went national and decisions got faster, the logic stayed human-shaped: you either paid your bills on time, or you didn't; you borrowed, or you didn't; you had a track record, or you were a question mark.

That older world had a certain blunt fairness to it. Not moral fairness—plenty of people were excluded for reasons that had nothing to do with responsibility—but mechanical fairness. A credit report was mostly a receipt book for debt. If you never used loans or credit cards, the system didn't call you virtuous; it called you *invisible*. And invisibility, in finance, is rarely a compliment.

Credit scoring industrialized that handshake. Instead of one underwriter's intuition, you got a number—an easy-to-sort, easy-to-price summary of your past borrowing behavior. Lenders loved it because it moved money at scale. Consumers learned to love it too, in a grudging way, because at least you could *see* the scoreboard. If your score was low, the reasons were often legible: maxed-out cards, missed payments, too many recent applications, too little history. The system still had flaws, but you could usually tell what it wanted from you.

Then the bargain started to change. Modern lending doesn't just want to know whether you pay *debt*. It wants to know whether you behave like someone who will pay *debt*. That sounds like the same question until you realize it invites a much bigger appetite for data. The new models aren't satisfied with "Did you repay?" They're curious about "How do you live?"—your financial rhythms, your stability, your routine, your risk, your friction, your predictability.

That shift is the difference between a handshake and a heatmap. A handshake is one signal: a moment, a decision, a yes/no. A heatmap is thousands of signals layered together until a pattern emerges. It's not personal in the warm sense; it's personal in the forensic sense.

The term you'll hear for the new inputs is *alternative data*: information that isn't part of the traditional credit bureau file but might help predict repayment. In practice, that can mean rent payments and utility bills, checking-account cash flow, payroll deposits, gig income, subscription patterns, device and location traces, shopping behavior, and a growing universe of "digital exhaust"—the data you leave behind just by living online. The promise is access: if you've never had a credit card, your on-time rent might finally count for something. The risk is that your life becomes a proxy test you never agreed to take.

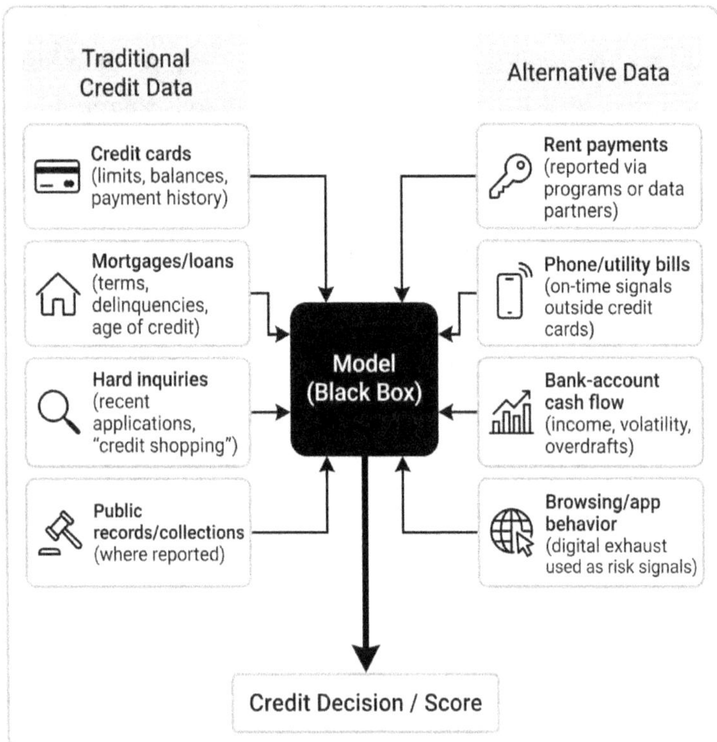

Here's the uncomfortable truth that makes alternative data so tempting: traditional credit scoring is excellent at predicting the behavior of people who already look like "credit people." If you have a few cards, a car loan, and a decade of history, the model has a rich record of how you handled exactly the kind of obligation a lender is considering. But if you're new to the country, wary of debt, paid mostly in cash, or simply never needed a loan, the file is thin. Thin files don't scream "responsible"; they whisper "unknown." And in lending, "unknown" often prices like "risky."

Alternative data tries to fill that silence with sound. If you pay rent on time for four years, that's not nothing. If your paycheck hits on a predictable schedule and your checking account doesn't lurch into overdraft every other week, that's meaningful. If you keep the lights on and the phone connected without drama, that suggests a kind of baseline stability. For someone who's been locked out of mainstream credit, this can feel like finally being allowed to bring evidence to court.

But evidence can be double-edged. Alternative data doesn't only add *good* information; it adds *more* information. And "more" is not automatically kinder. A model that sees your cash flow might notice that your income is irregular, even if your total annual income is fine. A model that sees your utility payments might note that you occasionally pay late—not because you're irresponsible, but because you're juggling. A model that sees your location patterns might infer instability from a long commute, a job change, or a move to care for family. The new system can be more inclusive and more intrusive at the same time, which is exactly why it's spreading.

To understand how AI changes credit, it helps to separate two questions that get tangled in everyday conversation:

- *What data is used?* (traditional bureau data vs. alternative data)

- *How is the decision made?* (simple formulas vs. machine-learning models)

You can have alternative data without fancy AI—just adding rent to a report. And you can have AI without exotic data—using machine learning to squeeze more predictive power out of the usual bureau variables. The real shift is that the industry is learning to do both at once: expand the inputs and upgrade the pattern-recognition engine.

Machine-learning models are good at picking up interactions humans don't naturally look for. A human might notice "high utilization is bad," meaning you're using a lot of your credit card limit. A model might notice that high utilization is *more* concerning when it rises quickly, *less* concerning when it's stable, and interacts differently depending on the age of the account, the presence of installment loans, or the pattern of payments. That kind of nuance can improve prediction. It can also produce decisions that are hard to explain in plain English, because the "reason" is distributed across dozens or hundreds of micro-signals.

That's where the heatmap metaphor earns its keep. Imagine two borrowers who both pay on time. Borrower A does it with a steady paycheck, a consistent checking balance, and spending that looks like a metronome. Borrower B does it with seasonal income, a few tight weeks each month, and occasional small overdrafts that get fixed quickly. Both are surviving. Both are paying. One looks like the system's favorite kind of boring; the other looks like noise. The new models aren't reading your morality. They're reading your volatility.

Volatility isn't a sin, but it can be expensive. Lenders price risk the way cities price waterfront property: the calmer and more predictable the view, the higher the demand. When AI models start to incorporate behavioral signals—how often your account balance dips, how frequently you change addresses, how regularly you pay recurring bills—they can penalize the financial version of weather. And weather is not evenly distributed.

There's another subtle shift: as lending becomes more automated, the "application" starts earlier than you think. In the old world, you applied, and then the lender looked at your data. In the new world, some systems pre-screen, pre-qualify, and market to you based on data trails that exist before you ever click "Apply." That can be convenient—instant offers, fewer forms. It can also mean you're being sorted long before you're listening, and the sorting can shape your options. If a model decides you're unlikely to qualify, you may never see the best offers in the first place. Denial becomes invisible, replaced by a quiet absence of opportunity.

The phrase "credit invisible" sounds almost poetic until you realize what it does to real life. It can mean paying a larger deposit to rent an apartment. It can mean relying on debit cards when you'd rather build rewards or protections that come with credit. It can mean higher insurance premiums in states where credit-based insurance scoring is used. It can mean a cell phone plan that costs more up front. In other words, it can mean paying more for being unknown—an ironic tax on people who haven't yet had the chance to prove themselves in the system that sets the prices.

So why not welcome alternative data with open arms? Sometimes you should. But only with your eyes open to three practical realities.

First, alternative data can behave like a "one-way mirror." You may feel like you're offering extra proof of responsibility, but you might not get to see what the lender sees. With traditional credit data, consumers have well-known rights to access their credit reports, dispute errors, and understand broad factors that drive scores. With newer pipelines—especially when data is pulled from bank accounts via permissioned connections, or from specialty databases, or from proprietary vendor models—visibility can get murkier. The less you can inspect, the harder it is to correct.

Second, more data creates more "gotchas." A single late credit card payment is obvious. A pattern the model interprets as distress might not be. Maybe you transfer money between accounts in a way that looks like "income smoothing." Maybe you front-load bills early in the month, creating a mid-month dip that resembles a cash crunch. Maybe your gig income arrives in irregular bursts that a model reads as instability rather than flexibility. A human might ask; a model just scores.

Third, the model's incentives are not the same as your incentives. You want a fair chance at affordable credit. The lender wants repayment with minimal loss. The model is trained to optimize the lender's objective, not to tell a story that feels fair to you. That doesn't make it evil; it makes it single-minded. When consumers feel wronged by automated credit decisions, the pain often comes from that mismatch: you experience a financial decision as a life decision. The model experiences your life as a data distribution.

None of this means the older system was kinder. "Handshakes" had their own ugliness: bias that lived comfortably behind closed doors, informal rules that punished outsiders, and subjective judgments that left no audit trail. One genuine advantage of algorithmic decisions is that they can be tested, monitored, and challenged—*if* the institutions running them bother to do that work, and *if* regulators demand it. A human can discriminate with a shrug. A model can discriminate at scale with a spreadsheet, which is why governance and accountability matter so much.

But even in a well-governed system, the consumer experience can feel surreal. You're not rejected by a person who can look you in the eye and say, "Here's what you should do next." You're rejected by a process that might give you reasons that sound like they came from a dropdown menu: "insufficient credit history," "high utilization," "too many inquiries." Those reasons can be true and still feel incomplete. The real driver might be a combination of signals no single sentence captures.

That creates a new kind of consumer skill: learning to manage not just your finances, but your *data shadow*. Not in a paranoid, tinfoil-hat way—in a practical, rent-is-due way. If lenders are increasingly reading stability, then stability becomes something you can design into your financial life, even if your income is messy.

Start with the easiest "heatmap wins": make the boring stuff boring.

If you can, put recurring bills on autopay—not because late fees are the main danger, but because regularity is a signal machines love. If autopay is too risky (because a surprise charge could trigger overdraft), set calendar reminders and pay manually, but aim for consistent timing. If your account balance often runs low mid-month, consider splitting bill payments across paychecks so your balance doesn't cliff-dive. You're not gaming the system; you're smoothing cash flow, which is good for you even if no one ever scores it.

Next, protect your checking account like it's a credit score. Overdrafts are more than fees; they're a story about stress. If you're frequently near zero, a small buffer can change your entire profile. That buffer doesn't have to be huge. Even one week of expenses parked in the account that pays bills can reduce the jaggedness that makes you look risky. The model can't know whether you're brilliant with money but underpaid. It can only see the jaggedness.

Then, be intentional about permissioning. Many modern credit products—especially fintech loans and "cash-flow underwriting" offers—ask you to link a bank account. That link can help you qualify, but it's also a data pipeline. Before you connect anything, ask yourself: am I doing this to get a better deal, or just to get *a* deal? If the offer is mediocre and the data access is broad, you're trading privacy for crumbs.

It also helps to recognize that the new system can punish frantic behavior. When money is tight, it's rational to apply for multiple options quickly, looking for relief. But many applications create inquiries, and inquiry clusters can drag down your profile in the short run. If you're shopping for a major loan, concentrate your rate-shopping in a tight window and use pre-qualification tools where possible. If you're applying for revolving credit because you're stretched, pause and consider whether the immediate credit line is worth the potential longer-term scoring hit.

One more uncomfortable point: alternative data can turn "life events" into "risk events." Moving frequently can look like instability. Changing jobs can look like volatility. Separating finances after a breakup can look like distress. None of these are moral failures; they're normal human transitions. If you're going through one, the best defense is documentation and consistency wherever you can create it: keep your address updated, keep bills in your name stable, keep one primary bank account steady, and avoid unnecessary account closures that erase history. In a system that values continuity, continuity becomes an asset.

All of this raises a philosophical question that becomes intensely practical the moment you're denied: should credit be about *who you are* or *what you've done*? Traditional scoring leaned heavily on what you've done with credit. Alternative data and behavioral models drift toward who you appear to be in motion—your patterns, your habits, your risk signature. That drift can expand access for people with thin credit files. It can also create a world where financial opportunity depends on how "machine-readable" your life is.

The takeaway is not "fear the algorithm." It's "treat the algorithm like a picky landlord." The landlord isn't impressed by your potential. They want proof that rent arrives on time, every time, with minimal drama. The new credit world is similar: it rewards routines.

Actionable takeaway: For the next 30 days, run a "stability sprint." Pick one checking account to be your financial home base, route income there, and pay your core bills (rent, utilities, phone, minimum debt payments) from that same account on consistent dates. Avoid overdrafts at all costs—reduce autopay if it risks surprise withdrawals, and instead set reminders you'll actually follow. The goal isn't perfection; it's a cleaner, calmer pattern that looks like reliability to both humans and models—and feels like breathing room to you.

6.2 Home Sweet Algorithm: Mortgages and Automated Valuations

The weirdest part of buying a home in 2026 isn't that you can tour a house via drone footage or sign documents from your couch. It's that the first "price" you meet often isn't a person's opinion—it's a number that shows up on your phone like a weather forecast. Sunny chance of appreciation. Lightly breezy comps. A cold front of "recent price reductions." You can feel your negotiating power change in real time, simply because an estimate exists.

That number has a job long before you do. It frames what sellers expect, what buyers offer, what agents pitch, and what friends text you ("Looks like it's worth $612k now!"). Even when everyone swears they're not relying on it, the estimate has already done the most powerful thing an algorithm can do in a market: it has made one version of reality feel *official*.

Mortgages used to be a tug-of-war between two human judgments: *Do we trust the borrower?* and *Is the house worth the money?* Modern mortgage pipelines increasingly answer both questions with models—credit models for you, valuation models for the property, and automated underwriting systems that stitch it together into a single yes/no gate. The result is an oddly emotional experience: you're trying to buy a home, but what you're really doing is trying to pass an invisible series of quality checks.

Start with the collateral side—the house itself—because it's where the "home sweet home" feeling collides with something colder: the idea that your dream needs to be *machine-readable*. Lenders don't just care that you love the place. They care that if you disappear (job loss, divorce, illness—life), they can sell the house and get their money back. The house is the lender's seatbelt. That's why the value matters so much, and why the value is increasingly calculated—at least in part—by an *Automated Valuation Model* (AVM).

An AVM is, in plain English, a house-price prediction machine. It takes data—recent sales, public records, property characteristics, sometimes neighborhood signals—and produces an estimate. Some AVMs are simple. Many are not. Modern versions can behave less like a spreadsheet and more like a pattern detector: it learns what "a three-bedroom in this zip code" usually sells for, then adjusts for features, lot size, renovations, market temperature, and the strange gravitational pull of nearby transactions.

Here's the part that catches people off guard: AVMs aren't just about replacing appraisers. They're about *controlling vari-*

ance. Human appraisals can differ, sometimes dramatically, because humans weigh comps differently, miss details, get influenced by presentation, or simply see neighborhoods through different lenses. Lenders hate uncertainty the way pilots hate turbulence. An AVM gives them a consistent, auditable process—especially valuable when loans are sold into the secondary market, where investors want standardized risk.

The secondary market matters because most mortgages don't stay where they're born. They're packaged, sold, guaranteed, securitized—a supply chain of housing finance that runs on standardization. In that world, a mortgage isn't only a relationship between you and a bank. It's a product that needs to meet a buyer's specs. Algorithms are how those specs get enforced at scale.

That's why the most important mortgage "decision" may happen before a human underwriter has your file open on a screen. If your loan is being evaluated for eligibility through an automated underwriting system (AUS), you're not just being compared to a policy manual. You're being compared to a statistical picture of what tends to go wrong.

Consider the FHA pathway, which is designed to expand access for borrowers who might not fit pristine, conventional profiles. FHA uses the TOTAL Mortgage Scorecard, an algorithm that classifies loans as "Accept" or "Refer". Those two words can change the entire tone of the transaction. "Accept" tends to mean the loan can move forward through an automated channel (with exceptions and possible manual downgrades). "Refer" often means a manual underwrite, more documentation, more scrutiny, more time—and more opportunities for the deal to wobble if the seller has other offers.

Notice what just happened: the algorithm didn't simply predict risk. It routed your life into a workflow. That's a big theme of algorithmic gatekeeping in modern finance. The model isn't merely a calculator; it's a traffic cop.

On the conventional side, government-sponsored enterprises (GSEs) like Freddie Mac and Fannie Mae sit at the center of the machine. Their automated underwriting systems don't just say "yes" or "no" in a vacuum; they provide findings and messages that shape what lenders ask you for and what they're willing to accept as proof. One underappreciated detail is that additional data can sometimes help—cash-flow assessment based on bank-account history, or rent payment history for first-time homebuyers, can be used to improve the credit-risk picture in certain workflows. That's the "credit invisible" idea moving from credit cards into mortgages: if you don't have a thick traditional file, the system may try to infer your stability from other signals.

This is the mortgage version of a modern resume scan. If you've ever applied for a job and suspected a bot rejected you before a person read your name, you already understand the vibe. The machine isn't offended by you. It's simply sorting.

Now layer the property side back in. In many transactions, the valuation process itself has become a stack:

- AVMs to estimate value quickly and consistently.
- Appraisals to provide human observation and justification, especially for higher-risk or higher-dollar cases.

- Collateral analytics tools to check whether the appraisal "makes sense" relative to data.

Fannie Mae's Collateral Underwriter, for example, is a collateral risk analysis tool that can flag appraisal risks by comparing the appraisal to underlying data and patterns. That's not just a second opinion—it's a *supervision layer*. The appraiser is no longer the final narrator of value; the narrative is audited by another model.

From a lender's perspective, this is risk management. From a buyer's perspective, it can feel like your home is being cross-examined by an unseen panel.

And then there's the interest rate—the part everyone feels in their stomach. Mortgage rates are influenced by broad market forces (Treasury yields, inflation expectations, central bank policy), but the rate *you* get is also shaped by how your loan is classified: credit score, down payment, debt-to-income ratio, loan-to-value ratio, property type, occupancy, documentation, and a host of eligibility rules. AI shows up here in two places: underwriting and pricing.

Underwriting decides whether the loan fits a box. Pricing decides how expensive that box is for you.

In a world of automated underwriting, underwriting and pricing have a tendency to merge. Instead of a human saying, "I don't love this, so we'll price it higher," the system may express caution by requiring mortgage insurance, limiting the program, demanding reserves, or routing you to manual review—each of which can increase cost, delay, or both. The algorithm doesn't need to "choose" a higher rate explicitly to make your mortgage more expensive; it can make your mortgage more expensive through requirements.

It's also easy to miss how property valuation influences pricing through the loan-to-value ratio (LTV). A small change in appraised value can move your LTV across a boundary that affects mortgage insurance, eligibility, or the interest rate adjustment applied. Two buyers with identical income and credit can end up with different monthly payments because one home's valuation cleared a threshold and the other didn't. In other words: you can do everything "right" as a borrower and still get priced by the house's data profile.

AVMs have their own quirks, and they tend to cluster around the kind of data problems that don't announce themselves politely:

- *Sparse comps*: rural areas, unique properties, or neighborhoods with low turnover can make model estimates swing wider.
- *Renovation blindness*: if improvements aren't captured in records, the model may undervalue what you can see with your eyes.
- *Neighborhood boundaries*: models learn patterns by geography, but real-life boundaries (a school district line, a busy road, a flood zone edge) can create sharp differences that models approximate imperfectly.
- *Fast-moving markets*: in a rapidly rising or falling market, yesterday's comps can be a liar.

None of these are exotic edge cases. They're normal features of housing. Housing is local, emotional, and heterogeneous—three traits that make prediction harder than it looks.

This is where regulation enters the room, not as a vibe but as plumbing. U.S. agencies adopted quality control standards for AVMs, with an effective date of October 1, 2025. The standards push institutions that use AVMs in certain housing-finance decisions to have policies and controls aimed at a high level of confidence in AVM estimates, protection against data manipulation, avoidance of conflicts of interest, random sample testing and reviews, and compliance with nondiscrimination laws. Translation: the government is treating property-valuation algorithms as critical infrastructure, the way you'd treat a bridge. You can't just eyeball it and hope it holds.

That "data manipulation" point deserves special attention because it reveals how the system thinks. An AVM is only as good as the data it ingests. If the data can be gamed, the valuation can be gamed. And in a world where valuations move money, people will try. The new standards are, among other things, an admission that algorithms create new attack surfaces.

There's also a consumer-side twist: the more valuation becomes algorithmic, the more a property's "digital twin" matters. The digital twin is the version of the house that exists in databases—square footage, bed/bath count, lot size, year built, permits, prior sale prices, tax assessments, and assorted descriptors. If your digital twin is wrong, the model can be wrong, and you can end up arguing with a number that feels strangely authoritative. The fight becomes less "This house is charming" and more "Your records say 1,620 square feet, but it's 1,820."

That's not hypothetical. It's one of the most practical ways buyers and sellers can influence outcomes: by ensuring the property data is accurate *before* it is scored, routed, or priced. Appraisals can correct some of this, but appraisals themselves are filtered through data expectations. When the algorithm expects one thing and the house is another, the transaction can slow down while humans reconcile reality with the dataset.

Now zoom out: AVMs and underwriting models are interacting with another algorithmic system most people forget is even there—*rental pricing algorithms*. In many cities, rents don't drift upward smoothly; they jump, stall, and jump again, influenced by software-driven price optimization. That changes the classic rent-vs-buy calculus in a subtle way. The old comparison assumed rent growth was mostly inflation plus a little local pressure. In an era of more dynamic pricing, rent can behave more like airfare: the same apartment can be "worth" different amounts depending on timing, vacancy, and the landlord's pricing model.

That matters for mortgages because "rent is the alternative." If the alternative gets more volatile, the emotional pressure to buy increases, even if the numbers don't support it. Algorithms can create a kind of synchronized squeeze: rising rents push people toward buying; valuations and underwriting tighten access; the people who least need help get through fastest.

Mortgage AI isn't only about who gets approved. It's about what kind of borrower you're encouraged to become.

If you want the cleanest algorithmic path, the system nudges you toward stability and documentation. Steady W-2 income reads better than lumpy gig income, even when the totals are similar. Cash reserves read better than "I could borrow from family if needed." A long, consistent address history reads better than frequent moves. A single, well-seasoned bank account reads better than a pile of accounts you open and close like browser tabs. The machine rewards legibility.

There's also a psychological trap hiding in "instant" decisions. Pre-approvals can feel like a verdict, but they're often conditional snapshots. They're based on the data provided, the moment in time, and the model's current rule set. A small change—new debt, a job change, a dip in cash balances, a property that appraises differently than expected—can flip the outcome later. The speed is real, but the certainty can be an illusion if you treat the algorithmic green light like a permanent passport.

So how do you live sanely in a mortgage world run by models?

One approach is to think like a quality assurance engineer. Not because you want to be robotic, but because the system is. Your goal is to reduce surprises.

A simple way to organize that is to separate what you can control into three buckets: borrower data, property data, and timing.

Bucket	*What "algorithm-friendly" looks like in practice*
Borrower data	Stable documented income; predictable bank balances; low revolving utilization; minimal new accounts close to application; clean explanations for any anomalies (job change, gaps, large deposits).
Property data	Accurate public records; documented renovations (permits where applicable); clear comparables; fewer "unique" features that are hard to price; awareness of appraisal/AVM risk in thin-comp areas.
Timing	Avoid major financial moves before closing (new car loan, furniture financing); understand rate locks; keep cash flow steady during underwriting; anticipate extra time for manual review routes.

The bucket most people ignore is property data, because it feels like the seller's responsibility. But if you're the buyer and the valuation comes in low, it becomes *your* problem fast. Suddenly you're renegotiating, bringing more cash, or walking away. A few hours of diligence can be worth thousands of dollars or an entire lost deal.

Ask early: What is the recorded square footage? Are there unpermitted additions? Are recent renovations documented? Are there comps that make sense, or is this the only modernized home on an otherwise untouched street? If the property is weird (and "weird" can be as simple as a converted garage), price in the possibility that a model will struggle. Models like patterns. Unique properties are anti-patterns.

On the borrower side, remember that underwriting models interpret behavior, not intention. If you plan to buy a home, the months leading up to it are not the time to run your financial life like a series of experiments. Avoid new credit. Keep balances stable. Keep documentation tidy. If you get paid in irregular ways, consolidate and label it clearly in your records so a human can make sense of it if the file gets routed to manual review. When humans step in, clarity is kindness.

Finally, treat mortgage pricing as an ecosystem. You can negotiate the home price, but you can also shop lenders, compare fees, and understand how small differences in points and closing costs translate into long-term expense. In algorithmic lending, different institutions may run similar models but interpret findings differently, or layer different overlays and pricing adjustments. Two lenders looking at the same borrower and same property can produce different outcomes—not because one is nicer, but because their systems have different appetites.

Actionable takeaway: Before you make an offer, run a "two-audit" check—one for you and one for the house. For you: pull your credit reports, keep revolving utilization low, and freeze major financial changes until after closing. For the house: verify the public-record basics (square footage, bed/bath, lot size), ask for documentation on major renovations, and flag anything unique that might confuse an AVM or trigger extra appraisal scrutiny. The goal is to make both borrower and property easy to underwrite, so you're negotiating with the seller—not with a surprise number that shows up at the worst possible moment.

6.3 The End of Shared Risk? Insurance Underwriting Reinvented

Insurance is one of the few financial products that feels like a small act of faith in strangers. You pay every month for something you hope never happens, trusting that if lightning hits your life—your house floods, your car gets totaled, your health takes a turn—the company will pay even though you're just one policy number in a sea of them. The only way that deal works is if the sea is big enough.

That "sea" has a name: *risk pooling*. It's the original hack that makes insurance possible. Lots of people chip in a little so that the few who get hit with a big loss don't drown. It's not charity; it's math with empathy baked into the structure. And for decades, the math was intentionally a bit blurry. Insurers grouped people into broad categories because the world was messy and data was limited, and because some amount of cross-subsidy is the price of having a functioning pool.

AI changes the blur. It doesn't merely refine categories; it dissolves them.

A simple way to see the shift is to imagine two neighbors, same age, same job, same car, same family size. In the old underwriting logic, those two people might look nearly identical: same zip code, same prior claims history (or none), same coverage limits. Their premiums would be close enough that no one would bother arguing about the difference.

In the new logic, insurers don't just ask, "Who are you, on paper?" They ask, "How do you *behave* in motion?" They don't only price the driver; they price the driving. They don't only price the homeowner; they price the micro-climate of the block. They don't only price the life insurance applicant; they price patterns that correlate with longevity, even when those patterns don't look like "medical information" in the everyday sense.

This is why the most important sentence about modern underwriting might be: *Insurance is becoming less like a shared safety net and more like a personalized invoice.*

The invoice can be accurate, in a statistical sense. It can also be socially explosive. Because the moment a pool becomes too personalized, it stops being a pool. It becomes a marketplace where everyone is priced as if they were their own tiny risk universe.

The Death of Pooling

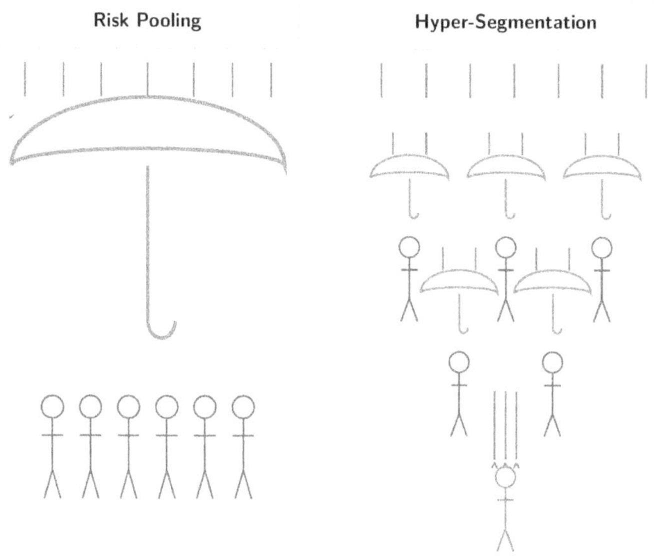

One umbrella, shared cost, shared protection

Many umbrellas, individualized pricing; some excluded

Pooling is intentionally imperfect. If insurance were a perfectly individualized product, your premium would be a mirror held up to your worst day. You would pay more precisely for the risks you bring, and you would receive more precisely the protections your risks deserve. That sounds *fair* until you remember what the word *insurance* is for. Insurance exists because life is lopsided. Some people get hit harder than others, and the timing is not politely scheduled.

A pool smooths that lopsidedness. It's a financial shock absorber.

The new underwriting playbook tries to reduce cross-subsidy by slicing the pool into thinner and thinner bands. That's what people mean by *segmentation*. The insurer isn't only separating "drivers with accidents" from "drivers without accidents." It's separating drivers who brake hard from drivers who don't, drivers who commute at night from drivers who don't, drivers who park on one block versus another, and eventually—if data and regulation allow—drivers who look like they're one bad month away from missing a payment from drivers who look like they have slack.

When segmentation becomes extremely granular, you get *hyper-segmentation*. The pool starts to look like a collection of little puddles. Each puddle is priced with more precision, which feels like efficiency. But puddles don't share. Puddles evaporate.

Why would insurers want that? Because cross-subsidy is expensive to the people doing the subsidizing. If you're low risk and you're stuck in a broad pool, you can feel like you're overpaying for someone else's losses. If an insurer can identify you as low risk with high confidence, it can offer you a lower premium and steal you away from competitors. That's not a moral argument; it's a competitive one. Markets reward companies that skim the safest customers.

This creates a pressure that's easy to describe and hard to escape: if one insurer gets better at identifying low-risk people, everyone else has to get better too, or they're left with a pool that's riskier on average. That's *adverse selection*—a classic insurance problem—now supercharged by modern data science.

The story sounds clean until you ask what "risk" actually means in the real world. A lot of risk is just physics: how far you drive, what the weather is, whether your house is in a wildfire zone. But a lot of risk is social structure wearing a math costume. Your zip code can reflect road quality, crime rates, emergency response times, flood infrastructure, building codes, environmental exposure, and—because history is not finished—patterns of segregation and disinvestment that didn't ask your permission.

So when an algorithm says, "People like you have more claims," the consumer's first question is reasonable: *People like me how?* Because "like me" can be a technical cluster (commute length, roof age, prior claims), or it can be a proxy for protected characteristics that the law treats as off-limits. Even when a model never uses race directly, it can pick up signals that correlate with race, income, disability, or other protected traits. That's the central tension: prediction can be accurate and still be unacceptable.

Life insurance makes the tension even sharper because the underwriting can reach beyond the doctor's office. Some regulators have been explicit about the risks of "external consumer data"—information not directly related to medical condition—being used to infer things that function as proxies for protected classes. The worry isn't science fiction. It's the ordinary creep of correlation: if a dataset predicts longevity and also predicts membership in a protected group, a model trained on it can quietly reproduce inequality while claiming it's just following the numbers.

The industry's favorite defense is also its most honest: "We're just pricing risk." But "risk" is not a neutral substance you scoop from a barrel. It's a decision about what to measure, what to count as relevant, and how to trade fairness against profitability.

This is where AI changes the feel of underwriting. A human underwriter can explain a decision in a narrative voice: "You have two speeding tickets and a recent at-fault accident, so the premium is higher." A machine-learning system can produce a rate that's correct by its internal logic while being difficult to translate into a story a consumer can recognize. The explanation can become generic because the true reason is a soup of tiny signals. And the signals can include things you would never think to defend.

That's why insurance regulators are increasingly focused on governance—how insurers manage models, test them, document them, and prevent outcomes that are arbitrary, capricious, or unfairly discriminatory. Model discipline, in other words, becomes consumer protection.

But governance doesn't change the core economic force pushing toward hyper-segmentation. It just tries to put guardrails on the road.

A practical way to understand what's happening is to replace the word "AI" with "microscope."

A microscope helps you see detail you couldn't see before. In underwriting, that means separating people who used to be grouped together. It also means discovering differences you didn't previously charge for. Some of those differences are legitimate and helpful. A good driver who rarely drives at risky hours may deserve a lower premium than a reckless driver with the same demographic profile.

The problem is that a microscope doesn't only reveal behavior. It can reveal *circumstance*. If you work a night shift, you don't drive at night because you enjoy danger; you drive at night because you're earning a living. If you live in a neighborhood with higher theft rates, you didn't personally vote for more theft; you're paying for a collective condition. Hypersegmentation can turn circumstance into surcharge.

And once the microscope exists, it starts to shape behavior. People begin to self-sort. Low-risk consumers chase low premiums and opt into programs that prove their low risk. Higher-risk consumers get priced out or pushed into thinner coverage. The pool gets less shared, not because anyone announced a new philosophy, but because the market mechanically rewarded the unsharing.

There's also a more subtle behavioral consequence: people start living for the score. Not a credit score this time, but an insurance score-sometimes literally, sometimes implicitly. If you know your premium depends on your driving patterns, you might brake more gently. Great. You might also avoid certain routes, avoid driving at certain times, or avoid reporting minor incidents for fear of being tagged as "claim-prone." Some of that is prudent. Some of it is anxiety management in the shadow of an algorithm.

A particularly sharp edge shows up when the model's feedback loop intersects with inequality. If high premiums force people into higher deductibles, they may avoid small preventative repairs. That can turn small losses into big claims later. The model then "learns" that these customers are expensive, reinforcing the premium hike. When pricing changes behavior, and behavior changes losses, the model doesn't just predict the future—it helps create it.

So who wins and who loses?

The winners are the people whose data trail reads as low variance. They don't just have fewer accidents; they have fewer surprises. They have stable commutes, stable housing, stable maintenance, stable cash flow. They are legible. They get rewarded for legibility.

The losers are not simply "high-risk people." They are often people whose lives are less predictable to the model: irregular schedules, multiple jobs, shared vehicles, crowded households, older homes, neighborhoods with infrastructure stress, gaps in documentation, thin files, complex medical histories, or simply bad luck stacked in the wrong year. Hyper-segmentation doesn't only punish danger; it punishes mess.

If this is sounding a bit bleak, it's worth saying out loud: there are real benefits to better risk measurement. When insurers can price more accurately, they can sometimes offer coverage to people who were previously rejected or given laughably high quotes. Better models can also detect fraud more effectively, which keeps premiums lower for everyone else. And behavior-based pricing can create incentives that reduce accidents and losses.

The question isn't whether prediction is useful. The question is what kind of social contract we want insurance to be.

A society can decide that certain risks should be broadly shared even if the data could individualize them. We already do this in many places, formally and informally. Some states limit or prohibit certain underwriting factors. Some programs exist precisely because private pricing would exclude too many people. Regulation is one way a society answers the question: *Should the pool be allowed to fragment this far?*

For an individual consumer, though, the decision is more immediate: *How do I avoid being mis-scored, mis-bucketed, or silently penalized by data I didn't know was in play?*

You can't personally reverse the economics of segmentation. You can, however, reduce the odds that you're paying extra for errors, outdated records, or lazy proxies.

One practical starting point is to treat your insurance profile like a file that can contain mistakes. Insurance companies use a variety of data sources and reports—claims histories, property databases, vehicle histories, and third-party risk scores. If your premium jumps, don't assume it's fate. Ask what changed. Sometimes the answer is boring: inflation, repair costs, catastrophe exposure. Sometimes it's not boring at all: a claim attributed to you that belongs to someone else, a property characteristic recorded incorrectly, a "risk" flag triggered by an address mismatch.

Another reality of the new underwriting world is that *shopping is now a form of due diligence*. Different insurers have different models, different appetites, and different ways of translating your data into a price. If one company decides you're expensive, it doesn't necessarily mean you are. It means you're expensive to *their* model. A second quote is not just about saving money; it's a second opinion on how the system is reading you.

Also, be deliberate about what you volunteer. Many insurers offer discounts in exchange for more data—apps, devices, monitoring programs. Those can be genuinely worthwhile for some households. They can also be a one-way door if you don't understand how the data is used, how long it's kept, whether it can affect not just discounts but surcharges, and what happens if you stop participating. "Discount" can be the friendly word for "data acquisition."

Finally, remember that underwriting is not only about price. It's about *access*. When the pool fragments, some people don't just pay more; they get offered less—higher deductibles, narrower coverage, more exclusions. If you're focused only on premium, you can accidentally accept a policy that looks affordable until the day you need it.

Actionable takeaway: The next time you renew (or get sticker shock on a quote), do a "pool check" before you just pay it. Ask what factors drove the change, request your claims history report if available, verify basic data points (address history, vehicle details, property characteristics), and get at least one competing quote to see whether the increase is "you" or the insurer's model. If you're offered a discount for sharing extra behavioral data, pause and treat it like a contract: understand what's collected, how it can change your price (up and down), and what happens if you opt out later. The goal is simple—stay insured without letting hyper-segmentation turn your life's messiness into a permanent surcharge.

6.4 Surveillance Pricing: When Your Car Snitches

A decade ago, "safe driver discount" meant something charmingly low-tech. Maybe your insurer trusted your clean record. Maybe your agent asked how far you commute. Maybe you swore, with the confidence of someone who never meets themselves in traffic, that you "rarely drive at night." The deal was simple: you promised you were low-risk, the insurer pretended to believe you, and everyone moved on.

Now the discount comes with a chaperone.

If you sign up for usage-based insurance—often shortened to *UBI*—you're agreeing to be priced not just on what you *are* (age, location, vehicle, claims history), but on what you *do* behind the wheel. UBI programs commonly collect telematics data: things like miles driven, speed, time of day, GPS location, hard braking, hard cornering, acceleration, and sometimes events like airbag deployment. "Telematics" is just a fancy word for "your vehicle (or phone) behaving like a sensor."

The pitch is seductively practical. If you drive less, you should pay less. If you drive smoothly, you should pay less. If you're a careful driver trapped in a world of chaotic drivers, why subsidize the chaos? It's the same logic that makes step counters popular: measurement feels like fairness.

But telematics turns measurement into a new kind of intimacy. Your insurer doesn't need to know your secrets to price you—yet location patterns have a way of becoming secrets all by themselves. A map of where you drive can imply where you work, where you worship, where you seek medical care, who you visit, what time you get home, and whether you're the sort of person who's out at 2 a.m. on a Tuesday. You don't have to confess anything. The pattern confesses for you.

This is the quiet pivot from "insurance" to "surveillance pricing." Not because every insurer is twirling a villain mustache, but because once a company can measure behavior cheaply, it becomes economically irrational *not* to. The discount isn't a gift; it's an invitation into a new pricing regime where your premium is a dashboard.

And the dashboard, like all dashboards, can be wrong in ways that matter.

One of the most unsettling consumer lessons of the past few years is that your car may be talking even when you're not listening. In January 2025, the Federal Trade Commission announced a complaint involving General Motors and OnStar over allegations that consumers were enrolled through a misleading process and that precise geolocation and driving behavior data was collected and sold without informed consent. Then, on January 16, 2026, the FTC finalized an order settling those allegations. The order included a five-year ban on GM sharing certain geolocation and driver behavior data with consumer reporting agencies, and it imposed long-lived requirements around affirmative express consent, consumer access and deletion requests, and options to opt out or disable certain collection (with limited exceptions, such as emergencies). The dates matter because they highlight something consumers often miss: this isn't a hypothetical risk. It's current-policy, modern-market behavior being litigated and regulated in real time.

Why consumer reporting agencies? Because once driving data enters the machinery of "consumer reports," it can influence not just a discount program you opted into, but broader insurance outcomes—premium changes, eligibility decisions, and the general sense that you're being judged by a file you've never seen. The terrifying part isn't that a safe driver might get a discount. The terrifying part is that a normal driver might get an *unexpected* surcharge because data took a detour through the wrong pipeline.

Even if you never drive a GM, the lesson generalizes: connected devices create data that can travel farther than you think.

Telematics also changes the emotional texture of insurance. Traditional underwriting felt like a once-a-year audit: you renew, you pay, you forget. Telematics can make pricing feel like a relationship with an opinionated roommate. Brake too hard and you can almost hear the app clearing its throat. Drive after midnight and it's as if the policy raises an eyebrow. Insurance becomes less about catastrophe and more about compliance.

Some people love that. It feels like control. It feels like, "Finally, my carefulness is rewarded."

Others feel an immediate tightness in the chest, because they recognize the trade hiding under the discount: you're swapping money for privacy, and privacy is hard to price until it's gone.

A $20 monthly discount feels concrete. "My insurer knows exactly where I drive" feels abstract. Abstract loses.

The catch is that the real "price" of telematics isn't only where you drive. It's what a model infers from how you drive. Hard braking, for example, can signal risky behavior—or it can signal living in a city where other drivers behave like they're being paid per lane change. Late-night driving can signal partying—or it can signal night shifts, second jobs, caregiving, or simply the time when your household is quiet enough to run errands. Speeding can be reckless—or it can be a short stretch where the speed limit drops unpredictably and the map app doesn't help.

Models don't understand your reasons. They understand correlations.

That's not a moral critique; it's a technical one. Telematics pricing is built to be scalable. Scalability means it can't interview you. It can't do nuance at the speed it wants to operate. So the question becomes: are the proxies it uses fair enough to live with?

The fairness question is not academic. It shows up in who can afford to opt in. If you're confident your data will look "good," the program is appealing. If your life is inherently messy—multiple jobs, long drives, irregular hours—you may expect the data to look "bad," which means you either pay more or you pay with privacy. That creates a new kind of sorting: privacy becomes a luxury good.

Telematics is also a gateway drug for broader behavioral underwriting. Once insurers get comfortable pricing driving behavior, it's not a long leap to pricing other behaviors—

especially when the data is already being collected by devices you wear.

Wearables bring the same logic to health and life insurance: your watch can track steps, heart rate, sleep, workouts, and sometimes more. The "wellness discount" pitch is almost too reasonable to argue with. If healthier behavior reduces claims, why not reward it? If reminders and incentives help people exercise, why not encourage them?

But wearables carry a uniquely intimate risk: *health data is not just personal, it's predictive.* A pattern of poor sleep, elevated resting heart rate, reduced activity, or irregular routines could be benign—or it could be an early signal of stress, depression, chronic illness, caregiving strain, or financial hardship. Once a model learns that certain patterns correlate with higher claims, it has an economic reason to price them, even if the consumer experiences them as temporary, situational, or private.

There's also the household problem. You might be comfortable sharing your own data, but what about everyone else who shares your car? A teen driver. A spouse. An elderly parent. A friend borrowing the vehicle. Telematics tends to treat "the car" as "the driver," which is convenient for pricing and awkward for fairness. The easiest way to get a bad driving score may be to have a life that involves other people.

And then there's the most mundane risk of all: *bugs*. Phone-based telematics can misread when you're a passenger. It can misinterpret subway movement, bumpy roads, or GPS drift. It can confuse "hard braking" with "pothole." A model can be statistically sophisticated and still be defeated by a cracked highway and a cheap phone mount. Consumers often assume the data is clean because it's digital. Digital is not the same as accurate.

So how should a normal person decide whether to opt in?

A good rule is to stop thinking of telematics as "a discount program" and start thinking of it as "a data contract." Contracts have terms, downsides, and exit conditions. Discounts are just candy at the front desk.

Before opting in, ask questions that sound boring but protect you:

- What specific data is collected (miles, speed, time, GPS location, braking, cornering, phone use)?
- Is the program *discount-only*, or can it also raise my rate at renewal?
- How long is the data retained, and can I request a copy or deletion?
- Can I pause or opt out, and what happens to my premium if I do?
- Is the data shared with third parties, and for what purposes?

Those questions aren't paranoia. They're basic consumer hygiene in a world where your premium may be downstream from a database you never meant to create.

Also, read the enrollment screens like you're installing a banking app, not a flashlight app. Look for the part where it asks for permissions—especially location permissions. "Always allow" is a different universe than "allow while using." If a program requires always-on precise location to function, that's not automatically wrong, but it should change how you value the discount. Always-on location is one of the most sensitive categories of data you can give away, because it's hard to make anonymous. A location trail is essentially a fingerprint.

If you do opt in, treat the first month like a test drive. Many programs show you a score or feedback. Watch for obvious errors (it thought you drove when you were on a train; it flagged you for phone use when the phone was in the glovebox; it recorded trips you didn't take). If the data looks sloppy, your price will eventually inherit that sloppiness.

If you're the kind of person who likes clean boundaries, consider device-based programs (where available) rather than phone-based ones, but don't assume device-based is automatically safer. "Safer" depends on what the device records, where it sends the data, and how it's governed. The question isn't "phone vs. device." It's "what's collected, what's shared, and what control do I have?"

Wearables require the same skepticism with a softer voice. Wellness programs can genuinely help people build habits. But they also create an incentive to perform health for the algorithm. That's fine when it means walking more. It's not fine when it means feeling punished for a week of insomnia, a depressive episode, a pregnancy, an injury, or a caregiving crisis. Your body is not a customer service ticket.

A final twist: surveillance pricing can change the meaning of "shopping around." In traditional insurance shopping, you compare premiums and coverage. With telematics, you're also comparing *measurement philosophies*. Different insurers weigh behaviors differently. One model may punish night driving more; another may emphasize mileage; another may treat city driving as inherently riskier because the braking patterns are harsher. You aren't only choosing a company. You're choosing a model's worldview.

That sounds dramatic until you remember what the model is doing: it's deciding whether your life looks like a claim waiting to happen.

The healthiest way to think about telematics and wearables is to treat them like any other financial optimization: worth considering, never worth surrendering yourself to.

The discount might be real. The trade might be worth it. But the decision should be deliberate, not default.

Actionable takeaway: If you're offered a telematics or wearable-based discount, run a two-step "surveillance pricing check" before you accept. Step one: demand clarity—write down what data is collected (especially GPS location), whether the program can ever increase your rate, and how to access/delete your data. Step two: run a 30-day pilot with your eyes open—monitor the app's recorded trips/behavior for obvious errors, and opt out early if the data quality looks questionable or the permissions feel too invasive. A discount is only a win if you still recognize the person being priced.

6.5 Instant Payouts and the ``Computer Says No" Problem

There's a specific kind of relief that comes from an insurance payout landing fast. Not "eventually." Not "after we assign an adjuster." Fast enough that it feels like the system noticed your emergency and decided not to turn it into a paperwork hobby.

That relief is real—and it's why automated claims processing is so seductive. A simple claim can be verified quickly, triaged cleanly, and paid in minutes or hours instead of weeks. For a renter whose laptop was stolen, or a driver with a cracked windshield, speed isn't a luxury. Speed is the difference between "annoying week" and "financial spiral."

But automation has a darker twin: the denial that arrives with the emotional warmth of a parking ticket.

The phrase "computer says no" isn't just a joke. It captures a modern kind of powerlessness: you're interacting with a decision that feels final, but doesn't feel *explainable*. There's no human voice to argue with, no obvious misunderstanding to correct, no shared reality to negotiate. There's simply a conclusion, delivered with the confidence of a system that has seen a thousand versions of you.

The uncomfortable truth is that claims are the moment insurance stops being a promise and becomes a test. You can pay premiums for years without learning much about your insurer's personality. Then you file a claim and meet the real product: the process that decides whether your story counts.

Automation makes that test faster. It doesn't necessarily make it kinder.

Why insurers automate claims in the first place

Claims are expensive even when the insurer pays nothing. Someone has to answer calls, request documents, read them, follow up, decide, record, comply, audit, and repeat. In many lines of insurance, the cost of handling the claim (the *loss adjustment expense*) can be a meaningful chunk of the premium you pay.

So insurers build machines that do three things at once:

- *Triage:* decide which claims are straightforward and which are complicated.
- *Verification:* check whether details match policy terms and known facts (coverage dates, deductibles, prior claims, police reports, provider billing codes, repair estimates).
- *Fraud detection:* flag patterns that look statistically unusual.

None of that is inherently sinister. In fact, it's what customers say they want: fewer delays, fewer forms, less waiting on hold. The hard part is that "unusual" is not the same as "fraudulent," and "statistically suspicious" is not the same as "wrong."

Models are good at detecting patterns. Life is good at producing exceptions.

Insurers also have a practical incentive to automate denials—not because denials are the goal, but because denial is a kind of control. If you can deny fast, you can stop the claim from becoming an open-ended expense. If you can delay, you can shift costs onto the claimant's patience and cash flow. Even when a denial is later reversed, the friction itself is a form of rationing.

That's why consumer anger tends to spike around denials, not premiums. Premiums hurt, but they're upfront and predictable. Denials feel like betrayal.

The dream scenario: three-second money

Automation isn't imaginary. Some companies have built highly streamlined workflows for small, low-complexity claims. Lemonade, for example, has publicly emphasized speed and automation in claims handling, while also stating it does not allow AI to take deterministic actions like rejecting claims or canceling policies. The fact that an automation-forward insurer feels the need to say that out loud is revealing: the industry knows exactly where public trust breaks. (lemonade.com)

In the best cases, automation is a consumer win:

- You report the loss in plain language.
- The system confirms your policy is active and the loss type is covered.
- It checks basic consistency (date, location, amount, prior claims).
- It pays, then closes the file.

Notice what makes this possible: *low ambiguity*. A broken phone. A stolen bike with a police report. A windshield crack. These are compact events. They don't require a philosophical debate about causation, medical necessity, or whether someone "should" have done something differently.

Automation thrives where the world is tidy.

The nightmare scenario: the algorithm finds you ``interesting''

The "computer says no" problem shows up when a claim is messy—when it requires judgment.

Sometimes "messy" is obvious: a major fire, a multi-car pileup, a long hospital stay. Those claims usually get routed to humans because the stakes are high and the evidence is complicated.

The dangerous category is the claim that *looks* simple but triggers a model's suspicion.

Fraud models are trained on history: known bad patterns, known good patterns, and the gray area in between. That history is inevitably incomplete. It can encode the industry's past biases about who "looks fraudulent" and what kinds of behavior "look suspicious." It can also encode structural realities—like the fact that people under financial stress may behave differently when filing claims (more urgency, less perfect documentation, more inconsistent records), even when they are entirely legitimate.

A model can't feel compassion for a person having a terrible month. It can only see the terrible month as a change in pattern.

And then there's the more bureaucratic form of "computer says no," where the denial isn't really about fraud. It's about rules, coding, criteria, and documentation. Health insurance is the purest example, because claims aren't about replacing a thing you lost. They're about deciding whether care was *medically necessary* under the plan's interpretation of coverage.

This is where the automation story stops being about convenience and starts being about power.

Health insurance's sharp edge: denials at scale

In April 2022, the U.S. Department of Health and Human Services Office of Inspector General (OIG) published findings that cut through the fog around denials in Medicare Advantage. Reviewing samples of denials from large Medicare Advantage Organizations, OIG found that among prior authorization requests that were denied, 13% met Medicare coverage rules—meaning those services likely would have been approved under Original Medicare. For payment requests denied, 18% met Medicare coverage and Medicare Advantage billing rules. (oig.hhs.gov)

Two things make those numbers unsettling.

First, they're not tiny. If you deny millions of requests, even a "small" improper-denial rate can represent a huge number of people stuck in limbo, forced to appeal, delay care, or pay out of pocket.

Second, the causes are painfully human, even inside a system that feels machine-driven. OIG pointed to denials linked to clinical criteria beyond Medicare coverage rules, "not enough documentation" even when records were sufficient, and payment denials caused largely by human error in manual reviews and system processing errors. (oig.hhs.gov)

That last phrase—*system processing errors*—should live rent-free in your mind. It means your claim can be "wrong" because someone didn't update a rule, or a system didn't map a code correctly, or a workflow dropped a document on the floor. The denial feels like judgment. The reality can be clerical.

When the denial is automated, the clerical mistake comes wearing armor.

Policy response: forcing reasons, forcing timelines

One reason the "computer says no" problem is so infuriating is that it combines speed with vagueness. A fast denial without a specific reason is not efficiency; it's a wall.

Regulators have been pushing back by demanding clearer explanations and more structured processes. CMS's Interoperability and Prior Authorization Final Rule (CMS-0057-F), released January 17, 2024, is part of that push. Among other things, it sets decision timeframes (72 hours for urgent requests and seven calendar days for standard requests) and requires specific reasons for denied prior authorization decisions beginning in 2026 for impacted payers. (cms.gov)

It also pushes the system toward APIs—standardized ways for providers and patients to exchange prior authorization information electronically—aiming to reduce the black-hole feeling where a request disappears and reappears as a denial with no clear path to fix it. Many API requirements have compliance dates beginning January 1, 2027 (with operational provisions generally starting January 1, 2026). (cms.gov)

This is the hidden point: regulators are increasingly treating denials as an information problem, not only a fairness problem. If you can't see why you were denied, you can't correct it. If you can't correct it, the denial becomes sticky, and sticky denials become a business model.

The consumer experience: speed that turns into whiplash

From the outside, modern claims automation creates a strange emotional whiplash:

- The app is friendly.
- The upload is easy.
- The status updates are instant.
- The denial arrives like a slap.

That's why people often describe automated denials as dehumanizing. It's not because humans are always nicer. Humans deny claims too. It's because humans can *participate* in the same reality you're in. They can acknowledge the details, ask clarifying questions, and sometimes admit ambiguity. Automated processes tend to feel like they already decided what kind of person you are.

The antidote isn't rage. The antidote is procedure.

When you appeal, you're not trying to "convince the algorithm." You're trying to force a claim out of the automated groove and into a review track where a human must reconcile your evidence with the denial reason.

Think of it as changing the venue.

How to fight an opaque denial (without losing your mind)

The biggest mistake people make after a denial is arguing the wrong thing. They argue fairness—"This is ridiculous!"—when the system responds to specificity—"This code was missing," "This clause excludes that damage," "This provider note didn't include X."

So the goal is to turn your claim into something the system can approve *within its own rules.*

Below is a practical appeal workflow that works across many kinds of insurance claims, with small adaptations.

Step 1: Freeze the facts while they're still warm

Write down, in one place, the timeline and the details:

- Date and time of loss or service
- What happened (two sentences, no embellishment)
- What you did immediately after (police report filed, mitigation steps, provider visit)
- What you're asking the insurer to pay and why

This isn't for the insurer at first. It's for you. Denials drag on, and memory decays under stress. A clean timeline is a superpower.

Step 2: Demand the actual denial reason, not the polite summary

A denial letter often contains a sentence that sounds like a reason ("not medically necessary," "insufficient documentation," "damage not covered"), and then—somewhere—references to policy language, codes, or criteria.

Your job is to locate the most *actionable* version of the reason:

- Which policy clause was applied?
- Which billing or service code was denied?
- Which documentation item was missing?
- Which clinical criterion was not met (health insurance)?

If the denial reason is vague, respond with a request that is hard to dodge: ask for the specific policy provision and the specific facts the insurer believes are true. Vagueness is the denial's camouflage.

Step 3: Build an ``evidence packet" like you're preparing a small trial

Appeals go better when you package evidence so a reviewer can say yes quickly.

A strong packet usually contains:

- A one-page cover letter with the timeline and what you want (approval, payment, reprocessing).
- Supporting documents labeled clearly (photos, receipts, repair estimates, provider notes).
- A short "map" paragraph: "Denial says X; attached document Y addresses X."

The point is to reduce reviewer labor. Automated systems are designed to minimize human time. If you want human time, you have to make it efficient.

Step 4: Fix the denial reason directly (even if it feels unfair)

If the denial says "insufficient documentation," the winning move is often not an argument; it's more documentation:

- For property claims: more photos, a contractor letter, proof of ownership, proof of pre-loss condition.
- For auto: repair shop supplements, police report, dash-cam footage, witness statements.
- For health: provider letter of medical necessity, clinical notes addressing criteria, corrected coding, prior authorization resubmission.

This can feel like playing along with nonsense. Sometimes it is. But appeals are rarely won by proving the insurer's process is annoying. They're won by making the file approvable.

Step 5: Escalate strategically

If the first appeal fails, escalation is about changing incentives:

- Request supervisor review or a different review unit.
- Ask for external review options if your plan/policy provides them (common in health coverage).
- File a complaint with the relevant regulator if the denial appears to violate clear rules or notice requirements.

Escalation works best when it's calm and documented. Anger is understandable. Documentation is effective.

Step 6: Track everything like a project manager

Create a simple log with date, time, person, and summary of each interaction. If you upload documents, save confirmation. If you mail, use certified mail when stakes are high.

This isn't obsessive. It's self-defense in a system built to be forgetful.

What to remember about ``instant"

Instant payout is a real feature. Instant denial is also a real feature. Both are powered by the same machine: pattern recognition plus workflow automation.

So the smart consumer stance is neither "trust the algorithm" nor "fear the algorithm." It's: *assume the first decision is provisional until you understand the reason.*

That mindset is especially important in health coverage, where prior authorization and claims denials can become barriers to medically necessary care. The OIG's 13% and 18% findings are a reminder that systems can be confidently wrong, and that appeals can reverse outcomes—sometimes because a human finally sees what the system missed. (oig.hhs.gov)

Actionable takeaway: If you ever receive a denial, do one thing within 24 hours: write a one-page timeline and request the specific denial basis in writing (policy clause, criteria, or code). Then assemble a labeled evidence packet that answers the denial reason directly and resubmit/appeal with a tracking log.

The goal is to force the claim out of the "computer says no" lane and into a review lane where a human can say yes.

Chapter 7

Planning for the Deep Future

Humans are notoriously bad at visualizing the future, and while AI excels at simulation, it cannot predict the regulatory or biological shifts of the next forty years. This chapter moves beyond the accumulation phase to the harder problem of 'decumulation'-how to spend your money without running out. We will explore why boring tax shelters remain your strongest defense against uncertainty, how to survive the specific mathematical danger of retiring during a crash, and how to treat ChatGPT as a research assistant rather than a fiduciary.

7.1 The Unchangeable Math of Tax Wrappers

Picture two coworkers, Maya and Chris, standing in the same hallway outside the same benefits meeting, each holding the same little plastic cup of lukewarm coffee. They're the same age. They earn about the same. They even invest in the same broad index fund, because they both decided—correctly—that trying to outsmart the market is a hobby, not a plan.

Thirty years later, one of them is quietly shocked by how much money is in the account. The other is quietly shocked by how *little* is left after taxes, even though the investments "did fine."

This isn't a story about genius stock picking. It's a story about *containers*. The boring-looking buckets your money sits inside—401(k), IRA, Roth IRA, HSA, taxable brokerage—aren't just filing choices. They're compounding machines with different gears. And once time gets involved, the bucket often matters more than what you pour into it.

The uncomfortable truth is that the market is hard to predict, but the tax code is *predictable in its shape*. Rates change, thresholds move, Congress tinkers. Yet the basic physics stay the same: when growth gets taxed along the way, your money drags a parachute behind it. When growth is sheltered, it runs clean.

Your portfolio has a silent partner: friction

Taxes are a form of friction. Not morally bad, not optional—just friction. And the reason friction matters is that investing returns don't arrive in one neat lump at the end. Returns drip in over time: dividends, interest, distributions, rebalancing gains, the occasional forced sale when you need cash. In a taxable account, each drip can trigger a tax bill. That money leaves the system forever. It can't compound for you anymore.

That steady leak has a name: *tax drag*. It's the performance you never see because it evaporates before the statement prints.

You can feel tax drag most clearly by comparing two versions of the same person:

- Version A invests in a taxable brokerage account and pays taxes along the way.

- Version B invests inside a tax-advantaged wrapper and keeps more money compounding internally.

Both could own the exact same index fund. Both could earn the exact same pre-tax market return. The difference is that Version A keeps skimming cream off the top for taxes every year, while Version B leaves the cream in the jar for decades.

And decades are where the math stops being polite.

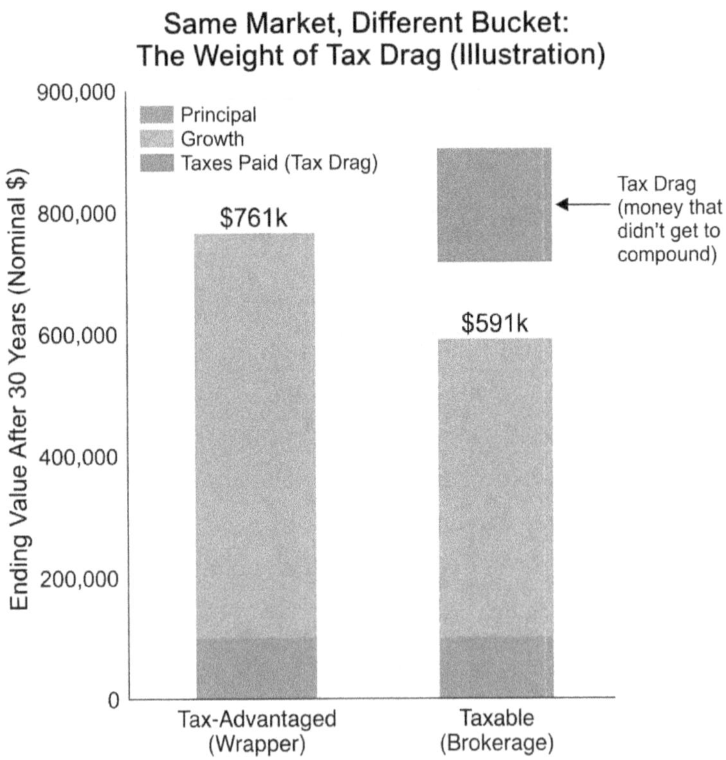

A small annual headwind doesn't look dramatic in year three. It looks dramatic in year thirty. That's the part our brains resist.

We're wired to fear cliffs, not slopes. Tax drag is a slope—an almost boring slope—until it's the reason two "similar" lives diverge.

Wrappers aren't investments. They're laws of motion.

A tax wrapper is not a stock. It's not a mutual fund. It's a set of rules that decides *when* taxes get assessed, *how* they get assessed, and sometimes whether they get assessed at all.

That sounds abstract until you translate it into a sentence your future self will care about:

> A wrapper decides whether your money compounds in a clean room or on a workbench full of sawdust.

The wrapper categories are conceptually simple:

- *Taxable*: you pay taxes along the way, and you may pay again when you sell at a gain.
- *Tax-deferred*: you generally dodge taxes during the growth years, but withdrawals are generally taxed later (think traditional 401(k) and traditional IRA).
- *Tax-free (under rules)*: you fund it with after-tax dollars, then qualified withdrawals can be tax-free (think Roth accounts).
- *The unicorn wrapper*: the HSA, which can offer tax benefits on the way in, during growth, and on qualified medical withdrawals.

Those are the "physics." The details-contribution limits, income phaseouts, which medical expenses qualify, which employer plan counts as compatible-are the knobs policymakers twist. But the underlying advantage is structural: *less tax friction over time usually means a larger pile at the end.*

That's why the wrapper decision is often a higher-return activity than arguing about whether a large-cap index fund should be "tilted" slightly toward value or quality. Those tilts might matter. But the wrapper is a lever that moves *every* year of compounding.

The most expensive sentence in investing: ``I'll do the account stuff later."

Many people treat tax-advantaged accounts like a paperwork chore: something you get around to when life settles down. The market doesn't reward that attitude, and neither does the tax code.

If you delay using wrappers, you don't just lose a one-time perk. You lose *years of sheltered compounding*. Years are not refundable.

It's like skipping sunscreen on the "not that sunny" days. One day doesn't matter. A pattern matters.

This is also where AI can be oddly dangerous. A slick app can help you choose funds in seconds, but it can distract you into thinking the fund choice is the main event. The wrapper choice is the main event. The wrapper is the stage lighting, the acoustics, and whether the building charges a cover every time you stand up.

A quick, human example: the ``same return" mirage

Suppose you and your sibling both invest $500 a month for decades. You're both disciplined. You both buy diversified, low-cost funds. You both average roughly the same market returns. You even rebalance occasionally, because you read the same article in the same year.

But one of you does it mostly in a 401(k) and Roth IRA. The other does it mostly in a taxable brokerage account because the 401(k) felt confusing and the login setup was annoying and, honestly, the phrase "beneficiary designation" sounded like something you'd need a notary for.

Years later, the brokerage investor is likely to have a surprising problem: the account is large, but the *after-tax* value feels smaller than expected. Selling to fund a house down payment or early retirement triggers capital gains. Dividends were taxed along the way. Rebalancing created taxable events. The whole thing worked-just with sand in the gears.

Meanwhile the wrapper investor is not "smarter." They simply ran the same engine with less friction.

This is why the phrase "my investments returned X%" can be misleading. The real question is: *what did you keep after the system took its cut, and when was the cut taken?*

The wrapper hierarchy: which bucket gets the ``best" dollars?

Here's the practical insight that feels almost unfair once you notice it: some dollars are more valuable than others, not because they're different dollars, but because of *where they're allowed to live.*

If you have access to a match in a 401(k), that match is usually the easiest money you'll ever see. It's not market alpha; it's payroll policy. Not taking it is like refusing a discount because the coupon has too many letters on it.

After that, the ordering depends on your situation-tax bracket now versus later, health costs, eligibility, and so on. But the *principle* is stable: fill the most powerful wrappers first, because they improve every future year of growth.

If you want a mental model that sticks, think of wrappers as layers of armor:

- The match is armor *plus* a boost.
- The HSA (if eligible and used strategically) can be armor in three directions.
- Roth-style space is armor against future tax rates.
- Traditional tax-deferred space is armor against current taxes and annual tax drag.
- Taxable accounts are still useful-flexible, no special rules for access-but they're the least armored.

A taxable brokerage account is not "bad." It's just honest. You get flexibility, and you pay for it with friction.

Why HSAs feel like a cheat code (and why they aren't effortless)

An HSA is often described as "triple tax-advantaged," and that phrase has launched a thousand personal finance rants. The reason people get evangelical is simple: the HSA can allow contributions that reduce taxable income, tax-sheltered growth, and tax-free withdrawals for qualified medical expenses.

But the HSA is not a magic loophole you tap once and forget. It's a rule-heavy wrapper. Eligibility depends on having the right kind of health plan. Withdrawals have to be for qualified expenses to get the best treatment. Recordkeeping matters if you reimburse yourself later. It's powerful precisely because it's narrow.

Still, if you're eligible and you can afford to pay current medical costs out of pocket, an HSA can become a stealth retirement account for healthcare-one of the largest, most stubborn expenses in later life. It's like pre-paying tomorrow's unavoidable bills with today's tax benefits.

The deeper point is not "everyone should max an HSA." The deeper point is that wrappers can be *purpose-built* for a category of risk you will almost certainly face. The wrapper is a policy tool aimed right at your future budget.

Traditional vs. Roth: the rate is a guess; the wrapper is still the lever

People get paralyzed by the traditional-versus Roth decision because it involves predicting future tax rates, future income, future policy, future everything. And yes: nobody knows. Anyone who tells you they know is either selling certainty or mistaking confidence for truth.

But you don't need perfect prediction to benefit from the wrapper logic.

Here's the cleanest way to think about it:

- Traditional accounts generally reward you when your tax rate today is higher than your tax rate in retirement.
- Roth accounts generally reward you when your tax rate today is lower than your tax rate in retirement.

That's the headline. The footnotes are where real life lives: Social Security taxation, required minimum distributions, Medicare premium brackets, state taxes, inheritance goals, career arcs, widow(er) tax brackets, and the possibility that retirement isn't a cliff but a long slope of part-time work, consulting, and weird little income streams.

So what do you do with uncertainty?

You diversify your *tax treatment*.

Tax diversification-having some money in taxable, some in tax-deferred, some in Roth-creates flexibility later. It lets you choose which bucket to pull from depending on the year. Flexibility is underrated because it doesn't show up as a crisp percentage return, but it can prevent expensive mistakes.

In other words: you don't need AI to tell you the "right" tax rate in 2056. You need a structure that won't fall apart if the rate is different than you hoped.

The stealth advantage of wrappers: they protect you from yourself

There's another wrapper benefit people don't like to admit: constraints can be a feature.

A taxable account is easy to raid. That's its advantage-and its hazard. A tax-advantaged retirement account comes with speed bumps: penalties, rules, paperwork, waiting. You can call that annoying. You can also call it protection against the version of you who decides, in a moment of stress, that your retirement money should become "temporarily" available.

Wrappers quietly shape behavior. They're not just tax rules; they're default settings.

This matters even more in an era of algorithmic persuasion. The future is going to have more personalized financial temptations, not fewer: "pre-approved" offers, one-click borrowing, buy-now-pay-later, targeted ads that seem to read your mind because, in a narrow sense, they do. A wrapper with friction-to-withdraw can be a surprisingly effective firewall.

Asset location: when the wrapper changes what you should hold

Once you accept that wrappers are different ecosystems, the next question becomes strangely practical: *what assets belong in which bucket?*

This is "asset location," which sounds like a term invented by someone who wanted to bill an hour, but the idea is intuitive:

- Assets that throw off a lot of taxable income (like taxable bond interest) often create more drag in a taxable account.
- Assets that are relatively tax-efficient (like broad stock index funds that don't distribute much) are often less painful in taxable.

- Assets with potentially high growth can be compelling in Roth-style space because the future growth can be sheltered.

None of this is a universal rule. It's a set of tendencies. Your time horizon, risk tolerance, and the actual funds you use matter. But the mental shift is the big win: you stop treating "portfolio" as one undifferentiated blob and start treating it as a set of compartments with different tax climates.

If you've ever wondered why two investors with the same asset allocation can end up with different after-tax outcomes, asset location is one of the biggest reasons.

The mistake AI will happily encourage: optimizing what's easy to measure

Algorithms love what's clean. They love what's labeled. They love what's downloadable as a tidy dataset.

Stock returns are tidy. Expense ratios are tidy. Backtests are tidy (even when they're misleading). Tax planning is messy-full of thresholds, exceptions, household details, and rules that change by jurisdiction and by year.

So a lot of "AI investing" will naturally steer people toward what's easiest to compute: portfolio optimization, factor tilts, rebalancing frequency, and tactical moves that look brilliant in a chart.

Meanwhile the biggest, most durable edge available to ordinary households is sitting right there with a name that makes your eyes glaze over: *tax-advantaged accounts*.

The danger isn't that AI will choose the wrong ETF. The danger is that it will make the ETF choice feel like the whole game-while the wrapper choice quietly dominates the long-run math.

If you remember one sentence, make it this: *The tax code is the house, and your portfolio lives inside it. A beautiful couch doesn't matter if the roof leaks.*

Actionable takeaway: run the ``bucket audit" before you chase better picks

Do this once, and then repeat it annually when you review your finances:

1. Write down every investing bucket you have: 401(k), 403(b), TSP, IRA, Roth IRA, HSA, taxable brokerage.
2. Next to each bucket, write two numbers: (a) how much went in last year, (b) how much *could* go in if you used it fully (based on your plan's rules and your eligibility).
3. Circle any employer match you are not capturing. Fix that first.
4. If you have an HSA and you're eligible to contribute, decide (explicitly) whether you're using it as a spending account, a long-term healthcare reserve, or not at all-then align contributions and recordkeeping with that decision.
5. Finally, choose a simple policy for tax diversification going forward (for example: "I will keep building both traditional and Roth space so future-me has options"), and

automate contributions so the plan doesn't depend on willpower.

If you do nothing else, do the automation. The best wrapper strategy in the world loses to the human habit of forgetting.

The market will do what it does. AI will suggest a thousand clever moves. Your job is to make sure your money is compounding in the right bucket while all that noise happens in the background.

7.2 Surviving the Descent: Withdrawal Strategies

Ellen retired on a Friday and woke up Monday with the weirdest problem: nothing had gone wrong, yet everything felt fragile. The spreadsheets were fine. The portfolio was fine. The plan was fine. But the paycheck-her old, steady, boring paycheck-was gone. In its place was a new job she didn't remember applying for: *portfolio manager of her own groceries*.

That first year of retirement is emotionally loud. Every expense starts to feel like a vote. Every market headline feels personal. And the usual investing advice-*stay the course, think long term*-suddenly sounds like it was written for someone who isn't trying to turn savings into a monthly life.

Saving money is a kind of uphill work. You can be wrong in small ways and still be okay. You can overpay for a fund, miss a month, panic a little and recover. Spending money down is different. It's not just "investing, but older." It's a separate math problem with sharper edges, because withdrawals force you to sell shares in real time, at whatever price the market offers that day, with no regard for your feelings.

The danger isn't that your investments underperform on average. The danger is that the market hits you with bad returns *early*, while you're also pulling money out. That combination can turn a perfectly reasonable lifetime of saving into a retirement that feels like it's always one bear market away from becoming a part-time job.

The risk nobody sees in a backtest screenshot

People love average returns because they're neat. They fit on one line. They make the future feel cooperative. But retirement doesn't live in the average; it lives in the *order*.

Sequence of returns risk is the plain-language name for a brutal truth: two portfolios can earn the same average return over 30 years, and one retiree ends up fine while the other runs out of money. Not because one chose better funds, but because one got unlucky with timing.

It's the same reason a strong swimmer can drown in a rip current. Strength matters. Conditions matter more.

Here's why the order matters. When you withdraw from a portfolio after a market drop, you're forced to sell more shares to raise the same amount of cash. That leaves fewer shares to participate in the eventual recovery. The market can bounce later and still not save you, because you already sold the future.

This is the part that feels unfair: in the accumulation years, volatility is mostly a mood problem. In the withdrawal years, volatility becomes an arithmetic problem.

A picture worth a thousand retirement calculators

The following chart is intentionally simple and intentionally upsetting. It shows two retirees starting with the same portfolio, taking the same annual withdrawals, and experiencing the same set of yearly returns-just in reverse order. Same average return. Same "long-term performance." Completely different outcomes.

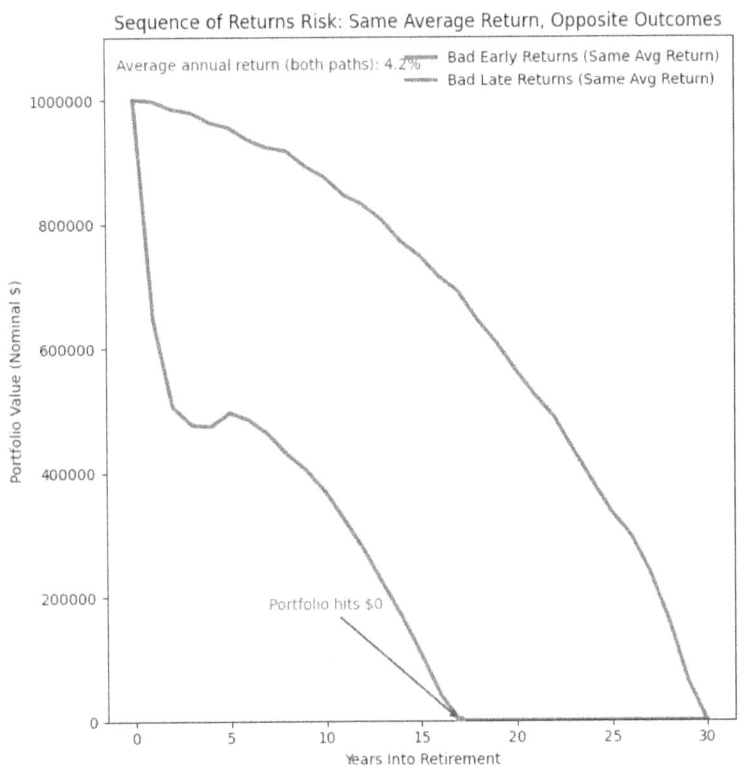

If you've ever felt guilty for caring about "when" you retire-"I shouldn't try to time the market"-this is the more compassionate reframing: you're not trying to time the market to get rich. You're trying to avoid retiring into a meat grinder.

You can do everything "right" and still step off the curb at the wrong moment.

Why the famous number (4%) is both helpful and dangerous

The 4% rule became famous for a reason: it's a clean starting point for a messy reality. The basic idea is simple: start by withdrawing about 4% of your portfolio in the first year of retirement, then increase that dollar amount with inflation each year.

The appeal is psychological. It turns a terrifying question ("How do I not run out of money?") into a single knob ("4%"). It gives you something to hold.

The danger is also psychological. People treat it like a law of nature, when it's really a historical observation wrapped in a rule of thumb. It came from looking backward at market history and asking, "What withdrawal rate would have survived across tough periods?" That's different from a guarantee that it will work for *you*, with *your* taxes, *your* spending, *your* retirement length, and *your* sequence.

Even when 4% is "safe" on paper, it can still be emotionally unsafe in practice. A plan can be mathematically survivable yet psychologically unbearable-especially if it asks you to keep spending the same inflation-adjusted amount during a market crash while your account balance is falling like an elevator.

If a strategy is so rigid that it makes you panic-sell at the worst moment, it isn't conservative. It's fragile.

Spending is not a bill. It's a control system.

Most retirement plans treat spending like a fixed bill: mortgage, groceries, travel, healthcare. But retirement spending is more like driving on an icy road. You don't set the steering wheel once and hope. You make small corrections. You pay attention. You adjust when conditions change.

A dynamic spending strategy is just that: spending that adapts. The point isn't to live in fear. The point is to treat spending as a lever you can pull *temporarily* to reduce the chance of permanent damage.

This is the core idea behind "guardrails" approaches: you set rules in advance that tell you when to cut spending and when you can safely raise it. Not because you love budgeting, but because you love not being forced into it later.

Think of guardrails as a pre-commitment device. You negotiate with yourself while you're calm, so you don't have to negotiate with yourself while markets are on fire.

There's a catch: dynamic strategies work only if the "dynamic" part actually happens. Many people love the theory and hate the moment. The math can tell you to cut spending 10% after a bad sequence; your life may already feel like it's cutting you 10%. This is where plans fail-not on spreadsheets, but in kitchens.

So the best dynamic strategy is the one you can execute without feeling like you're punishing yourself for being alive.

The retirement paycheck: build it from layers, not from hope

A practical way to make spending more resilient is to separate expenses into layers that behave differently in a downturn:

- *Must-pay*: housing, food, basic utilities, core insurance, baseline healthcare, essential transportation.
- *Nice-to-pay*: travel, gifts, major upgrades, generous dining out, the "we deserve it" category.
- *Optional and opportunistic*: big one-time spending that can wait for better markets.

This isn't a moral judgment about lattes. It's a mechanical tool. When the market drops early in retirement, you want to avoid selling more shares at depressed prices. The easiest way is to temporarily compress the discretionary layers while protecting the must-pay layer.

That's what successful retirees tend to do instinctively. They don't "follow the 4% rule." They behave like a household with a brain.

The cash wedge (or bond tent): boredom as a weapon

The most underrated withdrawal tool is also the least glamorous: holding a buffer of safer assets-cash and high-quality bonds-so you don't have to sell stocks during a market crash.

This goes by various names:

- a *cash wedge* (a stash of cash-like assets to cover near-term spending),

- a *bond tent* (a temporary tilt toward bonds around retirement to reduce early-sequence risk),
- or simply a *buffer*.

The point isn't to predict the bottom. The point is to avoid forced selling when prices are ugly.

A buffer is especially valuable in the first 5–10 years of retirement, when sequence risk is most lethal. It buys you time—time for the market to recover, time for spending cuts to be smaller and temporary, time for your plan to breathe.

There is a tradeoff. Holding more in low-return assets can reduce long-run growth if markets rise smoothly. But retirement isn't graded on perfect upside capture. It's graded on *not failing*. Buffers are insurance against the specific failure mode that ruins retirements: selling too much equity too cheap early on.

Boredom is the point. A buffer is the financial equivalent of keeping a flashlight in the drawer. You don't buy it because you're excited about flashlights. You buy it because the moment you need it, you really need it.

Wrappers come back with teeth: taxes and forced withdrawals change the descent

Withdrawals don't happen in a vacuum. They happen through accounts with rules-rules that can force your hand.

Tax-deferred retirement accounts (like traditional 401(k)s and traditional IRAs) often come with required minimum distributions (RMDs) later in life. Even if you don't *want* to withdraw that year, you may have to. That can push you into higher taxable income, increase taxes on Social Security benefits, and even increase Medicare premiums via income-related adjustments.

That's not a reason to avoid tax-deferred accounts; they're often fantastic. It's a reason to recognize that "withdraw 4% from the portfolio" is not a real instruction. The real instruction is: withdraw from *which account*, in *which order*, with *what tax impact*, while maintaining enough liquidity to avoid selling low.

This is where planning turns from elegant to real. A withdrawal plan is not just a percentage. It's choreography across buckets.

Roth accounts add a different kind of flexibility. Because qualified Roth withdrawals can be tax-free, Roth money can act like a shock absorber in years when taxable income is already high or markets are down and you want to avoid realizing gains. Taxable accounts add flexibility too-no early-withdrawal penalties, generally-but they can create capital gains and tax complexity.

So the wrapper choices you made while saving don't disappear at retirement. They become your tool belt. And the shape of that tool belt determines whether you can respond gracefully to a bad sequence or whether you're stuck selling what you don't want to sell.

The AI temptation: ``Just simulate a million futures"

Here's what a modern planning app can do impressively well: run Monte Carlo simulations, stress-test different allocations, model inflation, sample return distributions, incorporate longevity assumptions, and print a satisfying probability-of-success number in a clean font.

Here's what it cannot do: make you emotionally okay with cutting spending during a downturn, or magically prevent the first five years of your retirement from including a once-a-decade market event.

A probability is not a plan. A plan is what you do when the probability goes against you.

AI-powered tools can be genuinely useful here, but only if they're treated like a lab, not an oracle. Use them to answer questions like:

- "If I hold two years of expenses in cash, how does my failure risk change?"
- "If I agree to cut discretionary spending by 10% after a 20% portfolio drop, how does that change outcomes?"
- "What happens if inflation is higher for five years?"

Those are scenario questions. They lead to resilient behavior. What you want to avoid is the seductive output: "You have an 87% chance of success." Your life is not a casino bet you'll feel good about losing 13% of the time.

A better question is: *What would I do if the bad path happens early?* If the answer is "panic" or "go back to work," then the plan isn't finished.

A withdrawal plan that can survive contact with reality

A workable withdrawal strategy usually has three ingredients:

1. *A baseline spending target* that's realistic, not aspirational fantasy.
2. *A buffer* (cash wedge/bond tent) to reduce forced selling early in retirement.
3. *A pre-decided adjustment rule* so you don't debate yourself mid-crisis.

Notice what's missing: a belief that you can predict returns. You can't. That's not a moral failing; it's the market's job description.

And notice what's also missing: perfection. The goal isn't to find the "optimal" strategy. The goal is to find a strategy that keeps you from making irreversible mistakes under pressure.

Actionable takeaway: write your ``bad first five years'' script

Retirement planning gets dramatically more effective when you stop imagining the average future and start rehearsing the unpleasant one.

On paper (or in a notes app), write a one-page script that answers these questions:

- If the market drops 25% in the first two years, what expenses get cut first, and by how much?
- How many months (or years) of must-pay expenses do you want buffered in cash or high-quality bonds so you aren't forced to sell stocks at depressed prices?
- What is the specific trigger that causes a spending adjustment? (Example: "If the portfolio ends a year below X, spending next year drops by Y%.")

- Which accounts do you draw from first in a downturn, and which do you protect? (Even a rough order beats improvisation.)

Then do the most important part: tell someone you trust. A spouse, a partner, a friend, an adult child-someone who can remind you of your own script when headlines are screaming.

Because in retirement, the biggest risk isn't that your plan is imperfect. It's that a bad early sequence turns you into a different person for long enough to do permanent damage.

7.3 The Longevity Paradox and Healthcare Costs

Frank keeps a photo on his phone from the day he turned 65: a grocery-store cake, a grin that looks slightly forced, and a little handwritten sign that says "RETIREMENT." He took it because he wanted proof-proof that he made it to the age his father never reached.

Ten years later, the photo still makes him happy. It also makes him uneasy, because the same fact that makes it joyful is what makes it financially dangerous: he's still here, and he's healthy enough to want a long list of things. A longer life is the win. It's also the invoice.

Longevity turns retirement into a strange kind of planning problem. In most areas of life, "more time" is a gift you don't have to budget for. In retirement, time is the budget. Every extra year is another year of groceries, utilities, insurance premiums, dental surprises, a car that eventually becomes unreliable, and a body that slowly turns routine expenses into recurring ones. Investing can cover a lot of that. What investing can't do is make the timeline shorter or the healthcare system simpler.

The paradox is that we spend our working years trying to buy freedom-then discover that freedom comes with one expense category that refuses to behave like the others. Food inflation is annoying but somewhat predictable. Housing costs can often be stabilized with a paid-off mortgage or downsizing. Healthcare is different: it's both *steady* and *lumpy*. It is a monthly drip of premiums, copays, and prescriptions, and it is also a potential lightning strike that arrives late, expensive, and indifferent to your asset allocation.

If you're looking for the edge in planning for the deep future, it's not in predicting a particular stock's return in 2046. It's in admitting that a long retirement is not just "the same year repeated twenty-five times." It's a story with plot twists, and healthcare is the most likely author of those twists.

The hidden assumption inside most retirement math: a short old age

Many retirement projections quietly assume a certain shape of life: active early retirement, followed by a shorter period of decline, followed by the end. That shape is emotionally comforting. It's also not guaranteed.

A longer lifespan does something sneaky to the withdrawal problem. It doesn't just add extra years at the end; it adds extra years right in the middle-years where you're still living independently, still spending, still traveling a bit, still upgrading the house because your knees are now the decision-makers. Those "middle years" can be expensive in a way people don't expect, because they feel like regular life, not "old age."

And then there's the late-life period that people avoid imagining because imagining it feels like tempting fate: the years when health costs can spike and the question isn't "What do we want to do this summer?" but "Who is going to help with showers and medications?"

The market can handle long horizons. It was built for them. The problem is that humans tend to plan emotionally, and emotionally we plan as if the hard part will be brief. Longevity makes "brief" a risky bet.

Healthcare inflation isn't just inflation with a stethoscope

Most people can say something sensible about inflation. They've lived it. They've watched eggs and rent and insurance go up. But healthcare costs don't always rise like ordinary prices. They can rise because of prices, yes-but also because of *utilization* (you use more care as you age), *technology* (new treatments become available and expensive), *policy* (coverage rules change), and *pricing structures* that are famously hard to compare ahead of time.

This is why healthcare feels like a trap even for responsible planners. You can do everything "right," and then discover that "right" still includes writing checks you didn't practice writing.

It's not only the magnitude that matters. It's the unpredictability. People can adapt to expensive if it's steady. People struggle with expensive if it's jagged.

The two types of healthcare costs that wreck plans: the boring ones and the cinematic ones

Healthcare expenses tend to arrive in two storylines.

The first is boring: the steady monthly cost of being a person with a calendar. Premiums. Supplemental coverage. Drug plans. Dental cleanings that turn into dental work. Hearing aids that quietly cost the same as a used car. Physical therapy that becomes routine. A handful of prescriptions that eventually become a pharmacy relationship.

These costs don't blow up the plan in one dramatic moment. They slowly occupy more of the monthly budget until the retiree realizes the "fun money" category has become a rounding error.

The second storyline is cinematic: the hospital stay, the rehab period, the fall, the diagnosis, the surgery that's "covered" but still leaves you with a multi-thousand-dollar out-of-pocket bill, the in-home care arrangement that starts as "temporary" and becomes a new normal. The cinematic costs are what people fear. The boring costs are what people underestimate.

The truly dangerous years are when the two storylines overlap: you still have the boring monthly costs, and now you add a new high-cost need that doesn't respect your calendar or your portfolio's mood.

Long-term care: not a line item, a tail risk

Long-term care is where the longevity paradox gets sharp. In everyday language, long-term care means help with the activities of daily living-bathing, dressing, eating, mobility, supervision. It can be provided at home, in assisted living, or in a nursing facility. It can be part-time help that preserves independence, or it can be full-time care that becomes the household.

The reason long-term care is so hard to plan for is that it behaves like a tail risk: many people won't need years of expensive care, but the people who do can see costs that are brutally high, and those costs often arrive when the person is least able to earn, hustle, or "just tighten the belt."

It also tends to hit couples in a particularly cruel way. If one spouse needs care, the other spouse is often still alive-and still needs to live somewhere, still needs transportation, still needs their own medical care, still wants a life. The household's expenses can rise at the same moment the household's capacity to manage complexity collapses.

Markets are good at averaging outcomes across millions of investors. They are not good at ensuring that *your* household doesn't land in the expensive tail.

So the long-term care problem is not "How do I maximize returns?" It's "How do I prevent one category of risk from consuming everything else?"

Insurance is not an investment. It's a rent payment on stability

People often evaluate insurance like they evaluate stocks: "Did I win?" That mindset creates endless disappointment, because good insurance typically feels like a bad deal right up until the moment it becomes the best purchase you've ever made.

A better way to think about insurance is as a form of rent. You pay rent not because you hope the landlord returns it with interest. You pay rent because you want to control your downside. You want to live your life without a constant background hum of catastrophe.

Long-term care insurance (and newer hybrid products that blend life insurance with long-term care features) can be a way to transfer part of that tail risk to an insurer. Whether it's right depends on health, assets, risk tolerance, family support, and the local cost of care. It is absolutely not a one-size-fits-all choice, and it can be expensive and complex.

But the question to ask isn't "Will I get my money back?" The question is "What happens to my spouse and my choices if I don't have it and I need care for years?"

The emotionally honest answer is often what makes the decision clear.

The role of HSAs in the ``fourth quarter'' years

A Health Savings Account can look like a niche benefit until you frame it correctly: it is one of the few wrappers built specifically for a category of spending that tends to rise as you age.

The power of the HSA isn't just that it can help with next year's deductible. It's that it can become a dedicated pool for future healthcare costs-an account that can sit quietly for years, invested, waiting for the part of life when healthcare stops being occasional and becomes routine.

That framing matters because it changes how people behave. If you treat the HSA as a spending account, you'll swipe it like a debit card. If you treat it as a long-term healthcare reserve, you may choose to pay current expenses out of pocket (if you can afford to) and leave the HSA invested for later. The later-years version of you is often grateful.

This is not a call to hoard receipts or optimize your life into a compliance project. It's a call to match the wrapper to the risk. The risk is that healthcare becomes a dominant expense late in life. A dedicated bucket reduces the chance that healthcare costs force you to liquidate investments at the wrong time.

The most expensive retirement plan is the one that assumes a perfect family

When planners talk about long-term care, they often drift into polite phrases like "family support" and "aging in place." Underneath those phrases is the hardest variable in personal finance: people.

Some retirees have adult children who live nearby, have flexible schedules, and are emotionally and financially able to help. Some retirees have children who live across the country, are raising kids of their own, and are already stretched. Some retirees don't have children, or have complicated family relationships, or have children with special needs.

A plan that assumes "my family will figure it out" is not a plan. It's a hope-and hopes are fragile under stress.

This is where money, dignity, and control intersect. The ability to pay for help can preserve relationships. It can reduce resentment. It can let a spouse stay a spouse instead of becoming an exhausted nurse. It can let children remain children instead of turning into overworked project managers of appointments, paperwork, and medication schedules.

In other words: funding future care isn't only a financial choice. It's a choice about what kind of family dynamics you want under pressure.

The spending pattern that surprises people: early retirement can be expensive, and that's okay

Many people assume spending drops in retirement because commuting ends and the work wardrobe becomes irrelevant. Sometimes it does. But early retirement can also be a spending spike: travel, hobbies, home projects, helping kids, celebrating the freedom you worked to earn.

That spending pattern is not irresponsible. It's human. The mistake is building a plan that can survive only if spending is perfectly smooth.

A resilient plan anticipates that spending may be higher in the early years, lower in the middle, and then potentially higher again later because of healthcare. It also anticipates that healthcare costs can arrive even when you're "doing everything right."

When retirees get into trouble, it's often not because they spent too much on travel at 66. It's because they never built a realistic late-life healthcare scenario into the plan-and so the first serious health shock feels like betrayal rather than a known risk coming due.

Practical planning: build a healthcare ``shadow budget"

One of the best ways to make healthcare risk feel manageable is to stop treating it like a vague monster and start treating it like a set of categories you can name.

A "shadow budget" is not your daily budget. It's a parallel budget for healthcare and care-related costs that you keep separate on purpose. The point is psychological and tactical: you want to know what portion of your spending is non-negotiable and likely to rise, and you want to see it clearly.

A simple way to do this is to list healthcare-related expenses in plain language and give each a placeholder number (even if it's rough). Not because precision is possible, but because vagueness is expensive.

Category	*What it includes (examples)*
Premiums	Medicare-related premiums, supplemental policies, prescription plans
Out-of-pocket medical	Deductibles, copays, coinsurance, tests, specialist visits
Prescriptions	Ongoing medications, price changes, new medications over time
Dental/vision/hearing	Dental work, glasses, hearing aids, exams
Home support	Cleaning help, meal support, part-time aides, transportation help
Care contingency	A dedicated reserve for potential long-term care needs

Once those categories are visible, you can attach them to decisions you can actually make: whether to build an HSA reserve, whether to keep extra cash buffer in early retirement, whether to explore long-term care insurance options, and how aggressively you need to protect against early sequence-of-returns risk.

The shadow budget also makes it easier to have adult conversations with a partner. "We spend $X on healthcare" is less inflammatory than "You spend too much." It moves the discussion from blame to planning.

Where AI helps, and where it makes things worse

AI will be excellent at summarizing coverage options, explaining terminology, and helping you generate questions you didn't know to ask. It can translate insurance language into English. It can act like a tireless note-taker.

It will also be dangerously persuasive if you let it glide from explanation into prediction.

Healthcare planning is full of tempting questions that sound quantitative but aren't stable:

- "What will my healthcare cost at 87?"
- "How much will a nursing home cost in my area in 25 years?"
- "Which policy is best?"

A model can provide estimates. It cannot provide guarantees, and it can easily present guesswork with a confident tone. The risk is not just wrong numbers; it's false calm. False calm leads people to underinsure, under-save, or over-withdraw because a tool made the future look smooth.

Use AI as a question generator and a translator, not as a crystal ball. If it outputs a specific future number, treat that number as a placeholder and pressure-test it. Ask for ranges. Ask what assumptions drive the result. Ask what happens in the bad tail.

The most useful AI prompt for healthcare planning is not "tell me the cost." It's "tell me what could go wrong, and what I can do now to reduce the damage."

Actionable takeaway: a three-meeting plan with your future self

Set up three short "meetings" on your calendar, each with a concrete output. The goal is not to solve healthcare forever. The goal is to replace vague dread with specific, resilient moves.

1. *Meeting 1 (45 minutes): Build the healthcare shadow budget.* Create the categories above, estimate today's annual costs, and decide what you want to reserve for "care contingency" as a separate line. Output: one page.
2. *Meeting 2 (45 minutes): Choose the funding buckets.* Decide where the healthcare reserve will live (HSA if eligible, taxable account earmark, a conservative sleeve inside the portfolio). Output: a specific monthly transfer or contribution plan.
3. *Meeting 3 (60 minutes): Stress-test the tail risk.* Write down two scenarios: "one spouse needs part-time help for two years" and "one spouse needs full-time care for several years." For each, answer: who coordinates care, where the money comes from first, and what spending gets cut to protect the plan. Output: a written script you can hand to a partner or trusted person.

The point of all three meetings is the same: a longer life deserves a plan that doesn't collapse under the most predictable surprise in retirement—that bodies change, and the bills follow.

7.4 The AI Advisor: Fact-Checker or Fabricator?

On a quiet Tuesday night, Miguel typed a question into a chat box that looked friendly enough to babysit your dog: "Can I retire next year?"

The response arrived in seconds—smooth, confident, full of numbers. It read like competence. It felt like relief. And it was exactly the kind of answer a person wants when they're tired, anxious, and one news alert away from doom-scrolling themselves into insomnia.

Miguel wasn't reckless. He was doing what millions of smart people now do instinctively: asking the most available intelligence in the room. The problem is that the most available intelligence in the room is not always the most accountable.

Large Language Models—LLMs—are brilliant at producing sentences that resemble expertise. They can explain what a Roth IRA is in plain English. They can translate insurance jargon into human language. They can brainstorm tradeoffs you didn't think to consider. They can even sound like a calm, wise coach.

They can also make things up with a straight face.

That is the confidence trap: when an answer *sounds* precise, we treat it as precise. When it sounds tailored, we treat it as tailored. When it sounds calm, we treat it as true. The danger isn't that the model is "dumb." The danger is that it's eloquent in the exact way that causes people to stop verifying.

And financial planning is where verification isn't a nice-to-have. It's the entire job.

The bot's superpower: it can talk

An LLM's core skill is not math, and it's not judgment. Its core skill is producing plausible language based on patterns in the data it was trained on. That's why it can write a retirement explanation that feels clearer than the PDF you got from your plan provider. It can turn a messy question into a tidy narrative.

That narrative skill is incredibly useful—until you confuse it with a guarantee.

Humans have always been vulnerable to confident storytellers. For most of history, the confident storyteller was a charismatic neighbor, a talk-radio host, or a broker in a sharp suit. Now the confident storyteller is a clean interface with a blinking cursor. It doesn't interrupt you. It doesn't sigh. It doesn't judge your basic questions. It will answer the same thing ten different ways until it clicks.

So the temptation is understandable: why not use it as your advisor?

Because advisors are supposed to have duties. LLMs have *outputs*.

What an advisor is (legally) versus what a chatbot is (practically)

In the financial world, words like "advisor," "recommendation," and "fiduciary" are not just vibes. They are categories with rules attached. A regulated investment adviser is subject to legal obligations around client interests, disclosures, record-keeping, and oversight.

A general-purpose chatbot, even a very sophisticated one, is not automatically playing in that regulated category just because it speaks in the first person and uses the phrase "you should." It can feel personal while being fundamentally general. It can feel like advice while being closer to a well-written encyclopedia entry with improv skills.

This matters because financial decisions are asymmetric. If you ask a bot to explain a term and it's slightly wrong, you lose a few minutes. If you ask a bot whether to roll over a 401(k), claim Social Security early, convert to Roth, or cancel a policy, "slightly wrong" can cost you years of compounding, higher taxes, or a coverage gap you can't patch later.

The harm isn't that the model lies with malice. The harm is that it can be confidently incorrect without consequences.

The three failure modes you'll actually see

When people worry about AI in finance, they tend to imagine sci-fi catastrophes. The real hazards are more mundane—and therefore more likely.

First: hallucinated facts. An LLM can state a contribution limit, a phaseout range, an RMD rule, or a penalty percentage with a tone that implies it's reading directly from a government manual. Sometimes it is echoing something real. Sometimes it is echoing something outdated. Sometimes it is echoing something that was never true. You won't feel the difference from the writing alone.

Second: missing context. Your situation is not a generic example. Tax bracket interactions, state taxes, Medicare premiums, dependent coverage, business income, divorce, inheritance, disability, stock compensation—real life is an ecosystem. A model will often answer the question you asked, not the question your life is asking. It may not notice the landmines you didn't mention.

Third: false precision about the future. LLMs are great at describing what happened historically and what rules say today. They are not oracles. Yet they can produce a projected return, a "probability of success," or a "best" withdrawal rate with the same calm voice they use to define a deductible. If you don't stop it, it will happily turn uncertainty into a number because numbers feel like closure.

That last one is the most dangerous. Humans don't crave accuracy as much as we crave relief. A bot that gives you relief on demand is seductive.

A better mental model: use the bot like a research assistant, not a pilot

A good research assistant can be incredibly valuable. They can summarize, organize, draft, translate, and generate questions. They can save you hours of slogging through jargon.

A pilot makes decisions in real time when conditions change. A pilot is accountable for the landing.

In personal finance, you want the bot in the research-assistant seat. You want calculators, primary sources, and regulated professionals (when needed) in the pilot seat.

That division of labor is not anti-AI. It's pro-reality.

The ``two-source rule" for financial facts

One simple protocol lowers risk dramatically: never act on a financial "fact" you got from an LLM until you verify it in a second place, ideally a primary source.

Primary sources are the boring PDFs and official pages: IRS publications, plan documents, Social Security rules, Medicare documentation, your insurer's policy contract, your custodial statements. They're not fun. They're where the truth lives.

When you do this consistently, something surprising happens: the bot becomes *more* useful, not less. Because now it's not the authority—it's the translator.

A practical workflow looks like this:

- Ask the bot to explain the rule in plain English and list the exact terms you should search for in the official document.
- Look up the rule in the primary source.
- Bring the quoted language back and ask the bot to interpret what it means for a specific scenario, while you keep the source text as the anchor.

The bot can't bully you with confidence if you're holding the receipt.

The ``numbers test": make the bot show its work

When the question involves arithmetic—tax brackets, Roth conversions, RMDs, withdrawal rates, insurance premium comparisons—treat the bot's output like a student's answer on a math exam. If it can't show its work, it doesn't get credit.

Ask for:

- the exact assumptions (income, filing status, state, inflation, return),
- the formula used,
- and a sensitivity check ("What if the return is 2% lower?" "What if inflation is 1% higher?").

If the response becomes vague, contradictory, or overly certain, that's not a signal to try a different prompt until it tells you what you want. That's a signal to move the problem to a dedicated calculator or a professional.

A retirement plan that works only under one perfect set of assumptions is not a plan. It's a fantasy with footnotes.

The social-engineering problem: the bot may not be the bot

There's another AI risk that has nothing to do with retirement math: scams.

As AI-generated voices, videos, and emails improve, more fraud will arrive wearing familiar faces. A "quick call from your grandson." A "message from HR." A "note from your advisor." It will sound right, look right, and arrive at the exact moment you're busy.

This is where traditional financial hygiene becomes modern armor:

- Verify requests for money or account changes using a known contact method, not the one in the message.
- Use multifactor authentication.
- Create a family password for emergencies.
- Slow down any transaction that creates irreversible damage.

The most powerful feature of AI persuasion is speed. Your most powerful defense is friction.

When the bot is genuinely great

There are areas where an LLM can make you better at money, fast:

Vocabulary and translation. It can explain terms like "sequence of returns risk" or "coinsurance" without making you feel dumb. It can rewrite a plan document paragraph into plain English.

Question generation. It can help you produce the list of questions you should ask your plan provider, your insurance agent, or your CPA—especially questions you didn't know existed.

Scenario brainstorming. It can help you imagine "what could go wrong" and map out contingency plans: job loss timing, early retirement healthcare gaps, caring for a parent, widowhood, disability.

Drafting and organization. It can help write the email to your HR department, the letter to request a plan document, or the checklist you and your partner use for annual reviews.

Notice what these have in common: they're about clarity and preparation, not about irreversible execution.

When the bot is most dangerous

The danger zone is any decision that is:

- irreversible (or expensive to reverse),
- heavily tax-constrained,
- dependent on your exact personal details,
- or based on predicting the future.

These are the places where the confidence trap is most expensive. Roth conversions, claiming strategies, rolling accounts, choosing insurance coverage, selling concentrated stock, setting a withdrawal policy, deciding whether you "can retire"—these aren't trivia questions. They're life-shaping decisions with compliance edges and long tails.

A bot can help you prepare for those decisions. It should not be the thing that makes them.

A practical protocol: ``Ask, Verify, Execute"

If you want a simple rule that doesn't require you to become a regulatory expert, use a three-step protocol:

1. *Ask.* Use the bot to clarify concepts, list options, and translate jargon. Treat it as a friendly explainer.

2. *Verify.* Confirm any factual claims in primary sources or official calculators. Confirm any account-specific rules in your plan documents.
3. *Execute.* Make moves using reliable tools and channels: your custodian, your plan portal, your tax software, your professional advisor—whoever is accountable for the transaction and its compliance.

The difference between "Verify" and "Execute" is where people get hurt. They blur the steps because the bot sounds like it already did the verification. It didn't.

The soft problem: the bot doesn't know your fear

Retirement planning isn't only numbers. It's temperament.

Some people can tolerate a fluctuating portfolio and stick to a flexible withdrawal rule. Some people cannot, no matter how many simulations say they "should." Some people feel shame about spending. Some feel panic about being a burden. Some feel a deep need to leave money to kids. Some have experienced medical trauma that makes every healthcare decision emotionally charged.

A bot can ask you about "risk tolerance." It cannot see the way your hands shake when you log into your accounts. It cannot recognize the specific kind of exhaustion that makes a person click "confirm" just to stop thinking.

This is one of the best reasons to involve a human professional when the stakes are high: not because humans are always smarter, but because humans can notice the emotional context that turns a mathematically fine plan into a psychologically impossible one.

A note on humility: the best tool is the one that admits uncertainty

A strange rule of thumb: the more a financial answer sounds like it was delivered from a mountaintop, the more suspicious you should be.

Good planning is full of conditional language:

- "If you're in a higher bracket now..."
- "If your spending is flexible..."
- "If markets are poor early in retirement..."
- "If healthcare costs spike..."

A bot can be trained to sound humble, but humility in tone is not the same as humility in method. You want a process that forces uncertainty into the open: ranges, scenarios, sensitivity checks, and written contingencies. That's what keeps the plan from collapsing when reality refuses to be average.

Actionable takeaway: adopt a ``Can I Trust the Bot?" decision tree

When you're tempted to use AI for a financial decision, don't rely on your mood to decide whether it's safe. Use a repeatable rule. Put it somewhere you'll actually see it: a note on your phone, a printed sheet in your files, a screenshot pinned in your photos.

Here is a decision tree you can follow when the stakes are real.

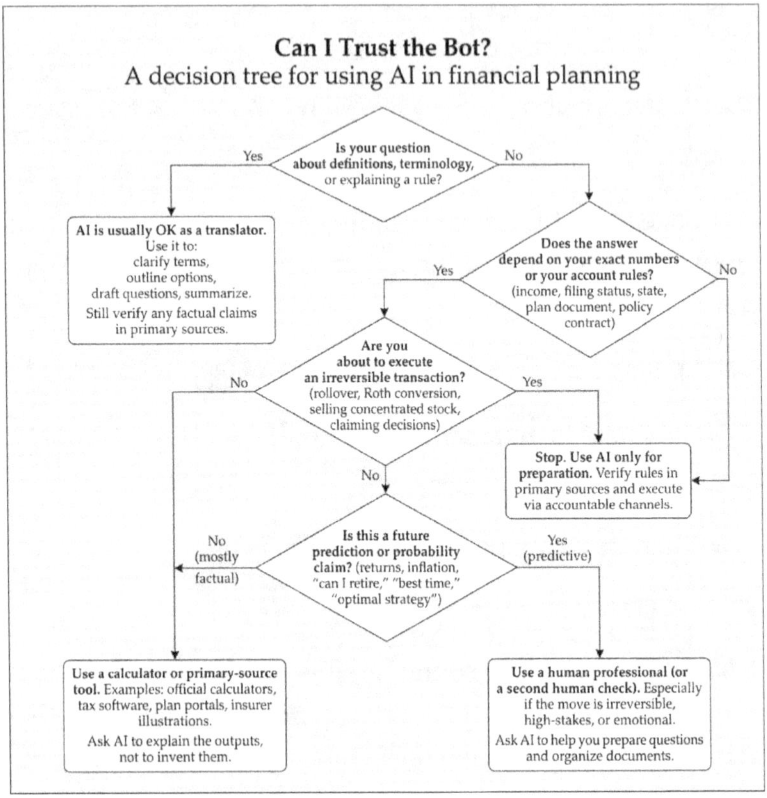

The point of the tree isn't to banish AI from your financial life. It's to keep you from outsourcing responsibility to a tool that can't be responsible.

Use the bot for what it's great at: clarity, language, brainstorming, and preparation. Use verification for what it's necessary for: truth. Use humans and calculators for what they're built for: accountable decisions.

The deepest future is unpredictable. That's exactly why your process can't be.

Chapter 8

The Human in the Loop

You wouldn't hand the keys to your car to a stranger without checking their license, yet we often hand our life savings to algorithms we barely understand. This chapter isn't about rejecting technology-it's about becoming the pilot rather than the passenger. We will explore how to build your own personal safety net, spot digital sleight of hand, and ensure that while AI does the heavy lifting, you remain the one calling the shots.

8.1 The CEO Mindset: Managing Your Personal Financial Stack

The first time a financial app gets something wrong, it rarely looks dramatic. It looks like a "helpful" category mistake—your rent tagged as "Entertainment." Or a cheerful notification that you're "on track," right before you remember you forgot quarterly taxes. Or a robo-portfolio that quietly drifts into a risk level you didn't mean to authorize, because you clicked through a questionnaire on a rushed Tuesday and never looked back.

Nothing explodes. No alarms blare. That's why it's dangerous.

Big financial institutions learned a long time ago that the scariest failures aren't always market crashes. The scariest failures are the ones that feel *reasonable* right up until they hurt you. A bank can have a beautiful model, a spotless spreadsheet, and a risk report formatted like a NASA launch checklist—and still lose money because the model had the wrong map of reality.

Households now have the same problem. Not because you're running a trading desk, but because your financial life is increasingly mediated by software that predicts, nudges, categorizes, approves, denies, and automates. You don't just have a budget and a brokerage account anymore. You have a *stack*.

And stacks need governance.

The two risks: the market, and the map

Most people understand *market risk*. It's the normal, headline-friendly kind of uncertainty: stocks drop, bond yields jump, your home value wobbles, inflation eats the purchasing power out of cash like termites in the walls. Market risk is the price of doing business in the real world. You can manage it—diversify, hold cash reserves, choose an asset allocation you can sleep with—but you can't delete it.

The new risk is sneakier: *model risk*.

Model risk is what happens when the math is right but the model is wrong.

That sentence deserves to be read twice, because it explains a lot of modern financial weirdness. The app can be doing exactly what it was designed to do—optimizing, classifying, recommending, forecasting—and still be wrong for *you* because its assumptions don't match your life. Or its training data doesn't match the current environment. Or its incentives don't match your goals. Or it can't see the thing that matters, because the thing that matters doesn't fit neatly into a database column.

Market risk is the weather. Model risk is the forecast model insisting it's sunny while you're standing in the rain.

A simple example: suppose your budgeting tool predicts you can "safely" spend $300 more this month because it sees an unusually low grocery bill. You know why the grocery bill is low—you've been traveling for work and eating reimbursed meals. The tool doesn't know that (or doesn't integrate it properly), so it interprets a temporary pattern as a new normal. The math is fine. The inference is not.

The CEO mindset starts with a blunt recognition: you are no longer managing money alone. You are managing money *plus tools*. Those tools have failure modes. And many of those failures won't announce themselves until after they've nudged you into a decision you didn't mean to make.

Your household is a small financial institution now

This is the part where people roll their eyes. "I'm not a bank." True. But look at your actual setup:

- Paychecks land in an account that might automatically distribute funds.
- Bills are paid by autopay rules you set once and forgot.
- Transfers happen through apps with tokenized access to your accounts.

- Credit decisions are influenced by automated underwriting.
- Investments may be rebalanced by an algorithm.
- Fraud detection is outsourced to machine learning systems you never see.

You may not have a risk committee, but you have a network of systems making decisions in your name. That network has dependencies: if one tool fails, another tool may act on the wrong information. That's not a theoretical problem. It's a very modern kind of fragility: when your financial life becomes a chain of "if-this-then-that," the weakest link gets to decide how your Tuesday goes.

Institutions don't handle this by finding the "best" model and trusting it forever. They handle it by building processes around models: oversight, limits, monitoring, and backup plans. Not because they're paranoid—because they've been embarrassed by reality enough times to develop healthy humility.

The personal version of that humility is simple: treat every important financial tool as a vendor. Vendors can be useful. Vendors can be wrong. Vendors can change their terms. Vendors can get hacked. Vendors can go out of business. Vendors can be acquired and quietly repurposed.

A CEO doesn't outsource accountability. They outsource tasks.

What, exactly, is your ``financial stack"?

Your financial stack is the set of apps, accounts, automations, and connectors that collectively run your financial life. Think of it like a modern kitchen. You might be a great cook, but if the stove thermostat lies, the fridge warms up, and the recipe app swaps teaspoons for tablespoons, dinner will be a tragedy even if your technique is flawless.

A typical stack has layers:

- *Core accounts:* bank(s), brokerage, retirement accounts, credit cards.
- *Decision engines:* robo-advisers, credit monitoring, "AI financial coach" tools, insurance pricing engines.
- *Automation rules:* bill pay, automatic savings, automatic investing, rebalancing, round-ups.
- *Data plumbing:* account aggregators and "connect your bank" permissions, exported transactions, linked email.
- *Interfaces:* the dashboards where you see your money (and therefore where your attention gets guided).

The CEO mindset doesn't require you to map this perfectly. It requires you to be able to answer, without guessing, three questions:

1. What systems can move money?
2. What systems can influence decisions?
3. What systems can see sensitive data?

If you can't answer those, you're not "bad with money." You're simply operating without an org chart.

Single points of failure: the most expensive convenience

Convenience is not free. It just charges you in a currency you don't notice: dependency.

The simplest single point of failure is obvious: having only one bank account, no backup card, and no emergency cash. But the modern single point of failure is often invisible: one app becomes the *source of truth* for everything—budgeting, bill pay, investing, identity monitoring, even tax estimates. If it breaks, your entire picture of reality breaks with it.

There's a special kind of stress that happens when you don't merely lose access to money-you lose access to *clarity*. The app is down. Your accounts don't sync. Your balances look wrong. Your transfers are "pending." Your autopay might still fire. You don't know what's real, and you can't make good decisions without reality.

Institutions try to prevent this with redundancy and separation of duties. You can do a household version:

- Keep at least one *independent* way to view each core account (direct bank login, not just aggregator view).
- Maintain a backup payment method (a second card, a small backup checking account).
- Keep a lightweight offline record of critical obligations: mortgage/rent, utilities, insurance premiums, tax dates.

That last item sounds quaint until you need it. The point isn't to live in a bunker. The point is to avoid being functionally helpless when a single service has a bad day.

Model risk in the real world: ``reasonable'' outputs that are wrong for you

Model risk doesn't always look like a scam. Often it looks like competence.

Here are a few ways it shows up in normal lives:

The budget that overfits your past. Many tools forecast your future based on your transaction history. That's fine until your life changes. New baby, new job, new city, caregiving for a parent, a medical condition, a divorce, a new commute—your spending pattern isn't just a curve; it's a cliff. Models love gradual change. Humans specialize in discontinuities.

A forecasting model may confidently "project" a savings rate that belonged to your old life. If you believe it, you'll feel like you're failing (or succeeding) for reasons that aren't real. The tool's confidence can become emotional leverage.

The investment profile that assumes your risk tolerance is stable. Many robo tools treat risk tolerance like eye color: answer a questionnaire, get sorted, done. But risk tolerance is more like sleep. It changes when you're stressed. It changes when your job is shaky. It changes when your partner loses income. It changes when you're staring at a down payment deadline.

A model can rebalance exactly as designed and still be wrong because it's honoring a version of you that no longer exists.

The "smart" categorization that gets the story wrong. Categorization errors aren't just cosmetic. They affect downstream decisions: alerts, spending limits, "safe to spend," subscription cancellation prompts, tax estimates, even your own perception of what you value.

If your childcare payments are misread as "Services," you may underestimate the largest financial project in your house: buying time. If business expenses are mixed with personal ones, the tool might quietly sabotage your tax planning. The model isn't malicious; it's just blind in the specific way it was built to be blind.

The optimization that optimizes the wrong thing. This is the adult version of "the GPS took me the shortest route through a swamp." Tools often optimize what they can measure: fees, volatility, cash drag, rewards points, "engagement," "net worth trend." Your life optimizes something messier: flexibility, sleep, family stability, the ability to say yes to an opportunity, the ability to absorb a hit.

A model may recommend maxing a certain account before building cash reserves because that's mathematically efficient under certain assumptions. If your household is fragile to cash-flow shocks, that "efficiency" is a liability disguised as sophistication.

The institutional trick: separate *decision* from *execution*

One of the most powerful controls institutions use is also one of the simplest: the person who decides is not the same system that executes without oversight.

In household terms, that means you don't let the same tool both *advise* and *move* without deliberate limits.

For example:

- A budgeting app can recommend a savings target, but it should not have unlimited ability to pull funds.
- A robo-adviser can rebalance within a portfolio, but you should be cautious about giving it wide permission to pull from your bank whenever it wants.
- An "AI coach" can draft a plan, but it shouldn't be the only place that plan exists.

This isn't anti-technology. It's pro-sanity. When advice and execution are fused, errors become expensive faster. When they're separated, you get a pause button.

A practical way to do this is to set *automation ceilings*. Instead of "invest whatever's left," set "invest $X on the 1st and $Y on the 15th," and revisit X and Y when your life changes. Or set thresholds where transfers require manual confirmation above a certain amount. The CEO mindset is full of these boring little guardrails—boring because they work.

Incentives: the question behind every recommendation

Institutions obsess over incentives because they've seen what happens when incentives quietly rewrite the mission.

Your tools have incentives too. They may be subtle, and they may be completely legal, but they matter:

- If an app makes money from order flow, it wants you to trade.
- If an app makes money from subscription, it wants you to feel dependent.
- If an app makes money from lending, it wants you to borrow.
- If an app makes money from interchange, it wants you to spend with *its* card.
- If an app makes money from referrals, it wants you to switch products frequently.

The model's outputs don't just reflect your data; they reflect the company's business model. If that feels cynical, good—that's your inner risk manager waking up.

A CEO-level habit is to translate every "recommendation" into a plain-language sentence that includes the incentive:

> "Based on what you told us, we recommend you do X"
> becomes
> "We make money when you do X, and the model believes X is defensible given the data we can see."

That doesn't mean X is wrong. It means you don't confuse a nudge with a fiduciary.

Diversify your dependencies, not just your assets

Most people know they should diversify investments. Fewer people think about diversifying *providers* and *access paths*.

Dependency diversification is not about having ten of everything. It's about not letting one company be the only key to your financial house.

A few CEO-minded moves:

- Keep at least one *secondary* bank relationship, even if it's just a low-activity account.
- Don't tie every account to the same email inbox without strong security and recovery options.
- Avoid making one app the only archive of statements, tax documents, or trade confirmations.
- If you use an aggregator, periodically verify balances directly at the institution—especially before big moves (wire transfers, home down payment, tax payments).

This is unglamorous. It is also the difference between "annoying inconvenience" and "full financial paralysis" when something goes sideways.

Governance for normal people: a lightweight risk department

A household risk department sounds ridiculous until you define it properly: it's a routine that makes sure your tools still match your life.

You don't need a binder. You need a rhythm.

Here's a governance structure that fits in real life:

1) The policy: what you will and won't automate. Write down (yes, write) a few lines:

- Which bills are always autopay.

- Which transfers are automatic and capped.
- Which actions must be manual (wires, large withdrawals, new beneficiaries, new linked accounts).

This is less about rules and more about clarity. When you're stressed, you default to the system you built. Build one you trust.

2) The inventory: what has access to what. Once a quarter, look at:

- linked bank accounts,
- connected apps/services,
- third-party permissions,
- devices logged in.

If this sounds like cybersecurity, that's because it is. But it's also finance, because access is a financial instrument now.

3) The limits: how bad can a mistake get before you notice? Institutions set limits because they assume mistakes will happen. You can do the same:

- Keep a buffer in checking so autopay doesn't cascade into fees.
- Set alerts for large transactions and low balances.
- Cap automatic transfers.

4) The audit trail: can you reconstruct what happened? If something breaks—an overdraft, a surprise fee, a misapplied payment—the most valuable thing you can have is a timeline. Ensure you can access:

- account statements directly from providers,
- transaction history exports,
- confirmation emails (or screenshots for major actions),
- customer support contact routes that don't rely on the app itself.

The CEO mindset treats "reconstructability" as a feature, not a luxury. When you can reconstruct events, you can dispute errors, correct patterns, and prevent repeats.

When the math is right but the world has changed

Model risk gets worse when the environment shifts, because models are trained on yesterday.

Sometimes yesterday is close enough. Sometimes it isn't.

Consider how many "normal" assumptions can break in a single year:

- Interest rates that were near-zero become meaningful.
- Inflation changes the meaning of "average" spending.
- Insurance pricing shifts as risk models update.
- A new fee structure changes the best account choice.
- A life change rewrites cash-flow stability.

A model that looked brilliant in one regime can look clumsy in another. That doesn't require a global crisis. It only requires drift: the slow accumulation of mismatch between the tool's assumptions and your reality.

This is why the CEO mindset is not "pick the best tool once." It is "assume drift, schedule check-ins."

A practical way to think: your money has three jobs

When you feel overwhelmed by tools, accounts, and competing advice, it helps to simplify what money is doing for you. Most households can map their finances to three jobs:

1. *Stability:* paying bills, avoiding catastrophe, staying liquid.
2. *Growth:* long-term investing, compounding, retirement.
3. *Optionality:* the ability to say yes—opportunities, moves, entrepreneurship, caregiving, rest.

Tools often focus on growth because it's quantifiable and exciting. Stability and optionality are harder to measure, which means models underweight them unless you deliberately insist.

A CEO doesn't let the dashboard choose the mission. They choose the mission and then decide what dashboards are allowed to optimize.

So when a tool recommends something, ask: which job is it serving? If it can't answer, or if it always answers "growth," you've found an incentive disguised as wisdom.

How to ``hire" tools like a CEO

CEOs don't just buy software. They hire it. Hiring implies:

- a job description,

- boundaries,
- performance expectations,
- probation,
- the possibility of termination.

Your job description might be: "Track spending, help me avoid overdrafts, and give me a clear picture of cash flow." If the tool starts trying to sell you lending products, it's doing a different job than you hired it for. If it requires broad permissions unrelated to its job, it's asking for too much access.

Probation matters because the first month is when tools reveal their true nature: how often they miscategorize, how confusing their fees are, how reliable syncing is, how hard it is to talk to a human when something breaks.

And termination needs to be emotionally easy. Tools try to become sticky—your history, your categories, your graphs, your identity verification, your "score." That stickiness is a business strategy. The CEO mindset says: portability matters. If you can't leave, you're not a customer. You're an asset.

Your personal Model Risk Register (yes, really)

Institutions keep "risk registers"—lists of risks, their likelihood, impact, and mitigations. You don't need an enterprise template. You need a napkin version.

Here's what it can look like in plain language:

Risk	*What it looks like*	*My mitigation*
Aggregator data wrong	App shows incorrect balances; duplicate transactions; missing accounts	Verify big decisions with direct login; keep cash buffer; don't rely on one dashboard
Automation cascades	One missed paycheck triggers overdraft + late fees + card decline	Low-balance alerts; autopay on credit card only for minimum; maintain backup account
Robo drift	Portfolio risk feels higher than intended; weird concentration	Quarterly review; rebalance boundaries; written target allocation
Model misunderstands my life	Tool thinks spending drop is permanent; recommends aggressive saving	Manually annotate unusual months; set savings based on plan, not forecasts
Tool vendor changes	Fees rise; new terms; acquisition changes product	Keep exports; maintain alternative provider; review permissions quarterly

The goal isn't bureaucracy. It's visibility. Once a risk has a name, it stops being a spooky feeling and becomes a manageable problem.

The CEO mindset in one week: an actionable takeaway

Pick one evening this week—twenty minutes, not a whole "finance day"—and do the household version of a board meeting. No spreadsheets required.

1. Write down your top five financial tools (apps, banks, brokerages, credit monitoring, robo-adviser).
2. Next to each, write three words: *moves money? sees data? gives advice?*
3. Circle anything that does more than one of those.
4. For every circled item, add one guardrail you can implement immediately (a transfer cap, an alert, a manual-confirm rule for large moves, a second way to log in, a direct-statement check before big transactions).

That's the shift: from "I use tools" to "I manage a system."

Markets will do what markets do. Your advantage isn't predicting them. Your advantage is building a financial stack that doesn't quietly make decisions *for* you while you're busy living your life.

8.2 Deepfakes and Digital Defense: Protecting Your Identity

Your phone rings at 9:47 p.m. Unknown number. The voice on the other end is familiar in the way a childhood smell is familiar—instant, bodily, undeniable.

"Grandma? It's me. I'm in trouble."

For a split second, your brain does what brains evolved to do: it tries to rescue the person you love. Not after a committee meeting. Not after a careful verification flowchart. Now.

That's the whole point.

Deepfakes aren't scary because they're clever. They're scary because they're *fast*. They bypass the part of you that checks facts and go straight for the part of you that protects family, avoids embarrassment, and wants problems to disappear. Fraud has always been a psychological sport; generative AI just gave scammers better equipment.

The uncomfortable truth is that "being good with money" now includes being good at one specific, unglamorous skill: refusing to act on urgency until you can prove reality is real.

The new scam advantage: it sounds like someone you trust

Old-school phishing often came with tells: awkward grammar, weird formatting, a "Dear Customer" greeting that made your eyes roll. Modern social engineering is smoother. AI can draft emails that read like your bank's compliance department wrote them on an espresso-fueled Tuesday. It can mimic brand tone. It can generate plausible transaction details. And, increasingly, it can mimic a voice well enough to make your gut react before your brain catches up.

Voice cloning changes the emotional math. An email asks for your attention; a familiar voice demands it.

The goal is rarely to steal your identity in a Hollywood way. The goal is usually simpler: get you to move money, reveal a code, approve a login, or "help" with a payment that you'll only question after the funds are gone. In many scams, the cash-out happens fast because the scammer knows your skepticism grows with every minute that passes.

So the defense has to be designed for speed too—not speed of reaction, but speed of *verification*.

A useful mental flip: treat every incoming message as untrusted input

There's a concept in security that translates perfectly into personal finance: don't treat an incoming message as truth just because it reached you. Treat it as *input* that needs to be validated.

This is surprisingly hard for humans because our social wiring assumes good faith. If a person sounds distressed, we want to help. If an email looks official, we want to comply. If a text says "Verify this login," we want the discomfort to end.

But the modern rule is blunt:

> Incoming calls, texts, emails, and DMs are not evidence. They are prompts.

Your job is to change the channel before you change the account.

That's what "out-of-band verification" means in plain English: don't keep the conversation inside the same potentially compromised channel. If someone calls claiming to be your bank, you don't keep talking and "confirm" details. You hang up and call the number on the back of your card. If someone texts claiming to be your kid, you call the known number you already have saved, or you use a separate family group chat.

It feels rude. It's not rude. It's competent.

The two attacks that matter most: account takeover and synthetic identity

Identity threats get described with a thousand terms, but most household damage comes from two buckets:

Account takeover. This is the classic "someone gets into an account that already exists"—your email, your bank login, your brokerage, your payment app. Once they're in, they can drain funds, change contact details, add external accounts, or simply watch quietly until the right moment.

Synthetic identity. This is weirder and, in some ways, more insulting: someone uses a mix of real and fake information to create a "new person" that credit systems accept. It might involve your Social Security number plus a different name and address. The goal isn't to impersonate you in full. The goal is to build credit under a manufactured identity and then cash out.

You don't need to memorize these terms to protect yourself. But you do need to know the practical implication: the most dangerous breaches aren't always about stealing money today. They're about gaining *a foothold*—a login, a code, a credit file—that lets an attacker move later with more power and less friction.

Your email is the master key (and most people guard it like a spare)

If you only upgrade security in one place, upgrade your email.

Email is the reset button for everything else. Password resets, "verify your new device," alerts, statements, and customer-service conversations all route through your inbox. If someone controls your email, they don't need to break into your bank with brute force. They can often stroll in through the "Forgot password?" door you didn't think about.

This is why a lot of financial fraud isn't a direct attack on a brokerage. It's an indirect attack on the identity layer around it: email, phone number, authentication codes, and customer support.

The practical conclusion is simple: your security budget—in time, attention, and money—should be concentrated where it reduces the most risk. That place is usually the email account tied to your financial life.

Passwords: stop trying to be clever, start trying to be boring

Passwords are a solved problem in the worst way: attackers have industrialized password cracking and credential stuffing. If you reuse passwords, you're effectively betting that none of the services you've ever used will ever leak. That's not a bet; it's a donation.

The modern "good enough" approach is boring by design:

- Use a password manager.

- Generate unique, long passwords for every important account.
- Stop using "security questions" as security (treat them as extra passwords and store the answers).

A password manager can feel like adding complexity. In reality, it's moving complexity out of your brain and into a vault that's built to hold it. The moment you stop relying on memory, you stop relying on habits. And habits are predictable.

Predictability is what attackers buy with automation.

MFA: not all two-factor is created equal

"Turn on two-factor authentication" is good advice in the same way "eat vegetables" is good advice: true, broad, and not specific enough to keep you out of trouble.

Multi-factor authentication (MFA) means proving you're you with more than one type of evidence. Usually:

- something you know (password),
- something you have (a phone, an authenticator app, a security key),
- something you are (biometrics like fingerprint or face).

MFA helps because a stolen password alone is no longer enough. That's the basic win.

But here's the modern twist: some MFA methods are much easier to trick you into handing over than others. If a scammer calls pretending to be your bank and says, "I just sent you a code—read it to me," they're not hacking your bank. They're hacking *you*.

And humans—tired, polite, startled humans—sometimes comply.

So the goal isn't just MFA. The goal is MFA that's harder to phish. In practice, the preference order looks like this for most people:

- *Best:* hardware security key or passkey-style login (designed to resist phishing).
- *Very good:* authenticator app generating codes (better than SMS, still phishable if you're tricked into typing the code into a fake site).
- *Better than nothing:* SMS codes (vulnerable to SIM swaps and social engineering, but still blocks many low-effort attacks).

If you only have the energy to improve one thing this month, improve the MFA on:

1. your primary email,
2. your mobile carrier account,
3. your bank and brokerage logins.

That ordering isn't about which account is "most valuable." It's about which account is the easiest path to all the others.

The SIM-swap problem: when your phone number becomes a vulnerability

A phone number feels like part of your identity, like an address. But in the security world, it's closer to a *routing instruction*. If an attacker convinces a mobile carrier to move your number to a new SIM, they can receive your texts and calls. If your bank relies on SMS codes, that can be enough to take over an account.

This is why some of the most financially expensive fraud stories start with a customer-service interaction at a carrier, not a sophisticated malware attack.

You don't need to become paranoid about carriers, but you do need to reduce how much power your phone number has:

- Move away from SMS MFA for critical accounts when you can.
- Add a carrier-level account PIN and keep it private.
- Be cautious with personal data that makes impersonation easier (full birthday, address history).

The theme is consistent: the attacker wants the easiest reset path. Your job is to make resets require you, not just information *about* you.

The ``safe word" is not cute; it's operational

Families tend to plan for fires and not for fraud. They teach kids what to do if the smoke alarm goes off. They don't teach kids what to do if an AI voice says "I need money right now."

A family safe word fixes that. It's a shared secret that's never used casually and never posted online. If someone calls sounding like your spouse or your parent and demands urgent action—wire transfer, gift cards, a Zelle payment, a crypto transfer—the safe word is the speed bump.

If they can't provide it, you treat the situation as unverified until you can confirm through a different channel.

This doesn't have to be a dramatic spy-code situation. It can be something mundane. The power comes from the fact that it creates a rule you can follow even when your emotions are hijacked.

People often say, "But what if I forget it?" Then write it down and store it like you'd store a spare key: not under the doormat, not in the open, but somewhere accessible when needed. The goal isn't perfect secrecy; it's protection against the most common, high-pressure impersonation play.

Your money-movement hierarchy: make irreversible actions harder

Not all payments are equal. Some are reversible, some are disputable, and some are the financial equivalent of tossing bills into a storm drain.

Scammers prefer payments that are:

- fast,
- hard to reverse,
- hard to trace.

So build a hierarchy:

- Make day-to-day spending easy (credit card with alerts).
- Make large transfers deliberate (manual confirmation, waiting periods, extra verification).
- Treat wires, gift cards, and crypto transfers as "break-glass" tools with an extra checkpoint.

This is not about fear. It's about designing friction into the actions that can ruin your month in thirty seconds. The CEO mindset applied to security is simple: add friction where mistakes are expensive; remove friction where behavior is safe.

The quiet superpower: alerts that tell you the story early

Most identity disasters are not single events. They're sequences:

- new device login,
- password reset,
- contact info change,
- adding an external transfer account,
- moving money.

If you only get alerted at the last step, you're already late.

So set alerts at the earlier steps. Many institutions allow notifications for:

- login from a new device,
- password changes,
- profile changes,

- large transactions,
- external account additions,
- address changes.

These alerts can be annoying. Good. Annoyance is cheap. Silence is expensive.

The key is to route alerts to a channel you will notice and that is reasonably protected. For many people, that means push notifications from the institution's app, plus email alerts to a hardened email account. If you're using SMS alerts, remember the phone-number risk and treat SMS as a convenience layer, not the only layer.

Open banking connections: the ``read-only" myth

Many budgeting apps and financial dashboards connect to your accounts through aggregators. The pitch sounds harmless: "We just want to read your transactions so we can help."

Sometimes it is read-only. Sometimes it isn't. And sometimes "read-only" still gives an attacker enough intelligence to do damage. If a criminal can see where you bank, what accounts you have, what your average balances are, and which bills you pay, they can craft a scam that feels eerily tailored. That's the point of modern social engineering: plausibility.

So when you link accounts, treat the connection like granting a visitor access to your filing cabinet. Even if they can't write checks, they can learn a lot.

A practical habit: once or twice a year, review connected apps inside your bank's settings (or the aggregator's settings) and revoke anything you no longer use. Old connections are the security equivalent of old house keys floating around in unknown drawers.

Customer support is part of the attack surface

A lot of people imagine hackers as keyboard wizards. In reality, the attacker's easiest target is often customer support.

If an attacker can convince support that they're you, they might:

- change your email address,
- change your phone number,
- reset authentication,
- bypass safeguards,
- add a new payee.

Companies try to prevent this, but prevention isn't perfect. The best defense you control is to tighten your own account recovery paths:

- Use strong MFA.
- Keep recovery email and phone number current.
- Avoid sharing personal details publicly that are used in verification (birthdays, address history).
- Consider a dedicated email for financial accounts if your main email is widely exposed.

It's not glamorous. It's the modern version of locking the back door.

What to do when something feels off: a micro-incident plan

People freeze during fraud because they don't know what "counts" as an emergency. The CEO mindset helps here too: define the first actions so you don't improvise while panicking.

If you suspect account takeover or fraud:

1. *Stop the bleeding:* lock or freeze the account if the institution provides that feature; change passwords from a clean device; log out other sessions if possible.
2. *Change the channel:* call the institution using a known number (card back, official website), not a link in a message.
3. *Secure the master key:* immediately secure your email account (password, MFA, recovery settings).
4. *Document:* note timestamps, transaction IDs, screenshots. You don't need a novel—just enough to reconstruct what happened.

Notice what isn't on that list: arguing with the scammer, replying to the text, clicking "unsubscribe," or trying to be polite. Politeness is a luxury item in a fraud event.

Credit freezes: the most underused control with the highest payoff

If you're thinking, "I don't have time to become a cybersecurity person," credit freezes are your friend. A freeze restricts access to your credit file, making it much harder for someone to open new credit in your name (or using your real identifiers in a synthetic identity scheme).

A freeze doesn't stop every kind of fraud, but it meaningfully raises the difficulty of new-account fraud. It's one of the closest things to a high-leverage move in identity defense.

The emotional barrier is that it sounds like a big step reserved for people who've already been victimized. It's not. It's like locking your car. You don't wait until it's stolen.

Unfreezing temporarily when you actually need new credit is usually manageable. The inconvenience is real. The protection is often worth it.

Deepfake-proofing your behavior: the ``slow is smooth" rule

Here's the psychological move that makes all the technical controls work better: slow down at the moment you're being rushed.

Scammers manufacture urgency because urgency collapses verification. A deepfake voice saying "don't tell Mom" is not just a story—it's a tactical instruction designed to isolate you. A fake banker saying "if we don't act now, the transfer will go through" is not just a warning—it's a way to keep you from hanging up and calling back.

So adopt one household rule that feels almost too simple:

> Any request involving money, codes, or account changes gets a call-back using a known number.

Not "most requests." Not "requests that seem suspicious." *Any request*. The power of the rule is that it removes the need to judge tone, plausibility, or emotion in the moment. You don't need to win an argument with a convincing voice. You only need to follow the rule.

This is the part people miss: security isn't primarily a technology problem. It's a behavior design problem. The best security habits are the ones you can execute while tired.

Actionable takeaway: the 20-minute identity hardening sprint

Set a timer for 20 minutes and do the highest-return moves—the ones that cut off the most common attack paths without turning your life into a compliance exercise.

1. Turn on MFA for your primary email account, ideally using an authenticator app or a phishing-resistant method if available.
2. Turn on MFA for your bank and brokerage accounts (upgrade away from SMS if you can).
3. Add login and profile-change alerts for your key financial accounts.
4. Create a family safe word and store it somewhere you can access in a real emergency.
5. Write one sentence on a note in your wallet or notes app: *"Call back using a known number before sending money or sharing codes."*

None of this makes you invincible. That's not the goal. The goal is to stop being the easiest target in the room—and to make sure that when a synthetic voice tries to hijack your emotions at 9:47 p.m., your system is strong enough to buy you the one thing scammers can't afford: time.

8.3 Know Your Rights: When the Algorithm Breaks

The most expensive lie in modern finance is a friendly sentence: *"Don't worry, we've got you."*

It shows up everywhere—on splash screens, in app-store descriptions, in the soothing tone of a chatbot that calls you "Alex" because it read your profile. The problem isn't that companies never help. The problem is that help is not a vibe. Help is a *legal obligation*, backed by a regulator, a rulebook, and a place to complain when the company fails.

When an algorithm breaks, the question isn't whether the interface looks professional. The question is: *what system are you actually in?* A bank system? A brokerage system? A payments system? A crypto system? Because your rights change dramatically depending on which bucket your money falls into—and modern apps are very good at blurring the buckets until something goes wrong.

The trick is to stop thinking like a consumer shopping for features and start thinking like an investigator labeling evidence. "Cash" is not just cash. "My investing app" is not just investing. "A debit card" is not just a piece of plastic. Each one lives under a different umbrella of protections, and the umbrella is not automatically extended just because the app uses familiar language.

Two kinds of failure: theft versus insolvency

When people say "I lost money," they often mean two very different things:

Someone took it. Fraud, account takeover, unauthorized transfers. The institution might be on the hook, but only if you follow certain procedures and timelines.

The company failed. The platform collapses, a partner bank fails, a broker becomes insolvent, a crypto exchange freezes withdrawals, a DeFi protocol gets drained. Here, the question becomes whether your asset was insured or segregated under a regulated framework.

Those two scenarios feel identical at 2 a.m. when you're staring at a missing balance. But the resolution path is completely different. One is about *error resolution*. The other is about *what your asset legally was* at the moment everything went dark.

AI makes this distinction more urgent, not less, because AI-powered interfaces can make every financial product look equally smooth, equally "smart," equally safe. A polished model can be the world's best camouflage for a weak legal structure.

Bank, brokerage, fintech: the labels that matter more than the logo

A traditional bank is boring in a very specific, very valuable way: it is heavily regulated, examined, and plugged into established consumer protection frameworks. That doesn't mean banks never fail or never make mistakes. It means there is a well-worn path for what happens next.

A brokerage is also regulated, but under a different set of rules. The core promise isn't that your investments can't lose value (they can). The promise is that customer assets are supposed to be handled in a particular way, and if the brokerage fails, there are mechanisms designed to protect customers from certain kinds of loss tied to the firm's insolvency.

A fintech app might sit on top of a bank or brokerage—or it might only *look* like it does. Many fintechs are essentially user interfaces plus marketing plus "data plumbing," with the regulated entity in the background. Sometimes that's fine. Sometimes it's the entire risk.

The CEO-style question to ask is blunt: *Who is actually holding my money, and what exact asset do I own?* Not "What does the app call it?" Not "What does the homepage imply?" The legal truth lives in account agreements, disclosures, and the identity of the underlying institution.

The false comfort of the word ``cash"

"Cash" is the most abused word in consumer finance.

In everyday language, cash means "the stable thing." In marketing language, cash means "don't worry, it won't behave like a stock." But legally and operationally, "cash" can mean:

- insured deposits at a bank,
- a money market mutual fund (not insured by deposit insurance),
- a brokerage sweep arrangement (may involve deposits at partner banks, with rules and limits),

- stored value in a payment app,
- a stablecoin (not a bank deposit),
- an IOU from a platform.

These are not interchangeable. Two products can both display "$10,000 cash" in the same font, with the same calming color palette, and yet have radically different protection in a crisis.

If you remember nothing else, remember this: in a failure, the interface disappears first. Your rights are whatever the underlying legal structure says they are.

The safety umbrella: what's actually protected, and what's not

The cleanest way to think about consumer protection is as an umbrella with holes: it covers some things well, covers other things partially, and leaves some things in the rain.

The biggest mistake people make is assuming that *familiarity* implies coverage. "It looks like a bank account" is not the same as "it is a bank deposit." "It's in an investing app" is not the same as "it's protected like brokerage assets." "It's called yield" is not the same as "it's interest on an insured deposit."

The moment you move into crypto, "AI yield farming," or anything that relies on smart contracts or loosely regulated intermediaries, the umbrella gets smaller fast.

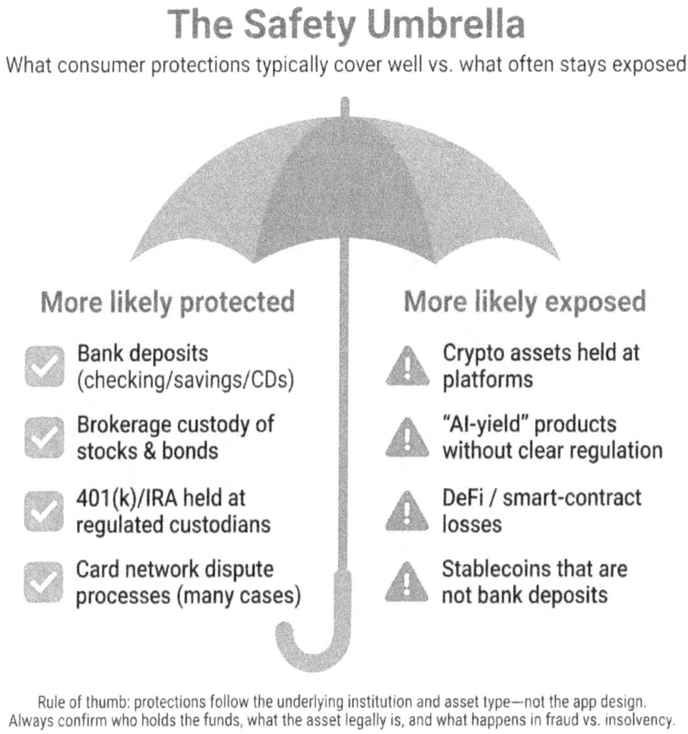

The point of that umbrella isn't to make you afraid of everything on the right. It's to keep you from confusing *new packaging* with *old protections*. Plenty of innovation is legitimate. Plenty of it is useful. But when something breaks, the resolution often depends on whether you're standing on a regulated floor—or on something that only felt like a floor because the UI was convincing.

If money moves electronically and it's wrong: you have a process (and deadlines)

When an unauthorized electronic transfer hits your account—whether it's a debit, an ACH transfer, or another kind of electronic movement—the most important thing you can do is *act like there is a clock*, because there is.

Consumer protections for electronic fund transfer errors generally come with procedures: you notify the financial institution, it investigates, it must communicate results, and in many cases it may need to provisionally credit you if the investigation takes longer than certain time limits.

Here's the part people don't do because it feels "too formal": they don't create a paper trail. They call, get put on hold, feel embarrassed, hang up, and hope it resolves itself.

A stronger approach is operational:

- Report the error immediately through the institution's official channel (phone number on the card, official app, official website).
- Write down the date and time, the person (or case number), and exactly what you reported.
- Follow up in writing if asked (and even if not, consider doing it anyway if the situation is serious).

This is not about being dramatic. It's about being legible to a system. Algorithms may flag the problem; humans may resolve it; but the machinery of resolution runs on documentation.

A useful mindset: treat unauthorized transfers like a kitchen fire. Don't debate whether it's a "real" emergency; just turn off the gas, call the right number, and document what happened.

Debit card versus credit card: the ``whose money is it right now?" test

A credit card is often safer for daily spending than a debit card for one simple reason: when fraud happens on a credit card, the thief is stealing the card issuer's money first, not your rent money.

With a debit card, fraud can drain your checking account directly, and even if you're eventually made whole, the timing can be brutal. Bounced payments, overdraft fees, cascading failures of autopay—this is where a "small" fraud event becomes a week-long financial migraine.

This isn't a moral argument about debt. It's a cash-flow argument about shock absorption. Using a credit card for routine spending (paid off in full) can act like a buffer layer between fraud and your ability to function.

And it pairs neatly with the identity defense habits from earlier: alerts, strong authentication, and the discipline to call back through known channels before you "confirm" anything.

Credit decisions by algorithm: the right to a real reason

AI shows up in lending in a way that can feel personal: you apply for credit and the answer comes back cold, fast, and confident. Denied. Or approved, but at a rate that feels like a judgment on your character.

A key consumer right here is deceptively simple: when a creditor takes an adverse action (like denying your application or offering less favorable terms), you're generally entitled to a notice that provides *specific reasons*.

The reason this matters in an AI world is that "the model says no" is not a reason you can use. A vague explanation is not just frustrating; it's useless. You can't improve what you can't understand, and you can't dispute what you can't name.

So if you get an adverse action notice that feels like a shrug—something generic like "failed to meet internal standards"—treat that as a signal that the system may not be respecting the spirit of the rules. Your leverage often starts with a simple act: saving the notice and asking for clarity through the right channel.

That channel is not the chatbot.

When your credit report is wrong: dispute like you're building a case file

Credit reporting errors can feel like trying to argue with a ghost. The information is wrong, you didn't do the thing, and yet the consequences show up in real life: higher insurance premiums, worse loan terms, or a denial that arrives like a slammed door.

Disputing an error works best when you treat it like a case file, not a complaint:

- Identify each specific item you're disputing.
- Explain why it's wrong in plain language.

- Attach documents that support your claim.
- Keep copies of everything you send.

It's tempting to write a passionate paragraph about how unfair it feels. Save that energy. Systems don't process vibes; they process claims tied to evidence. The "CEO mindset" applied here is not corporate coldness. It's a way of being kind to your future self: making it easy to escalate if you need to.

And if the error is connected to identity theft, don't improvise a recovery plan from scratch. Use the official, structured resources designed for that scenario so you're not reinventing a process in the middle of stress.

Identity theft recovery: the difference between panic and a plan

Identity theft is one of the few financial experiences that can make competent adults feel instantly incompetent. You start questioning your own memory: "Did I open that account?" You start second-guessing every click you've made for the last six months. You start looking for a single "undo" button that doesn't exist.

The best counterweight to that panic is a step-by-step recovery plan. A good plan does two things:

- It reduces the number of decisions you have to make while emotionally flooded.
- It creates a consistent narrative you can hand to banks, bureaus, and agencies.

The practical move is to use official recovery tools that generate a checklist, help you track progress, and create documentation you can reuse. This isn't about bureaucracy. It's about conserving your attention for the few actions that actually matter: securing accounts, freezing credit if needed, disputing fraudulent accounts, and documenting everything.

Scammers rely on chaos. A plan is the opposite of chaos.

The complaint ladder: don't argue with the app

When something goes wrong, many people get trapped at the lowest, stickiest level of the system: in-app support.

You will be invited to chat. You will be given scripted empathy. You will be asked to reboot, reinstall, re-authenticate, and "rest assured." Sometimes that's fine. Sometimes it's a time sink that quietly burns through the very thing you can't replace: the window in which timely reporting strengthens your position.

A more powerful approach is to know the complaint ladder:

- Start with the institution that holds the account (not the glossy interface).
- Escalate through formal channels when the stakes justify it (written disputes, supervisor escalation, secure messages).
- Use regulators and consumer complaint portals when appropriate, especially if you're being stonewalled.

This is not about being adversarial for sport. It's about matching the seriousness of your problem to the seriousness of your channel. A missing "$12" subscription charge is a chat issue. An unauthorized "$8,000" transfer is a formal-process issue.

The fintech trap: ``partner banks" and the mirage of borrowed trust

Many fintechs advertise trust by association: "We work with an FDIC-insured bank." That statement can be true and still misunderstandable.

The important questions are:

- Is *your* money actually held as a deposit at the insured bank, in a structure that qualifies for deposit insurance?
- Or is your "balance" something else—stored value, a claim on the fintech, a sweep you didn't fully understand, or a pooled structure with conditions?

This isn't an accusation. It's a reminder that financial plumbing is complicated, and marketing copy is rarely written by the plumbing department.

If you can't quickly figure out the underlying institution and the nature of the asset, treat the product like it's in the right-hand "exposed" side of the umbrella until proven otherwise. That simple default can prevent a lot of regret.

The crypto reality check: code doesn't owe you customer service

Crypto and DeFi often come with a philosophical pitch: no middlemen, no gatekeepers, pure self-custody, pure freedom.

The fine print of that freedom is brutal: if a smart contract fails, if you send funds to the wrong address, if a protocol is exploited, there may be no regulated party required to fix it. There may be no error-resolution timeline. There may be no one to "reverse" anything. Sometimes there isn't even a phone number to call—just a Discord server and a community manager typing condolences.

Even when a platform is centralized and looks like a fintech app, the assets involved can fall outside the protections people assume from traditional finance. You might have terms of service, but not the kind of insurance framework people mentally associate with a bank.

This doesn't mean no one should ever touch crypto. It means the decision should be made with eyes open: you are stepping out from under one umbrella and into a different weather system.

The AI twist: a black box is not a legal loophole

When an institution uses an AI model for decisions—credit, fraud flags, account closures, risk scoring—it might feel like the machine has the final say. It doesn't.

A core idea in consumer protection is that complexity does not erase obligations. If the law requires specific reasons for a credit denial, "the model is complicated" is not a free pass. If error-resolution procedures apply, automation doesn't nullify timelines. If a firm delegates part of a process to a vendor, it doesn't delegate away responsibility.

The practical implication is empowering: when you hit a wall of "our system won't let us," that may be a workflow problem, not a rights problem. Your job is not to reverse-engineer their model. Your job is to insist—politely, persistently, and in the right channels—on the procedures you're entitled to.

Actionable takeaway: the 12-minute rights check before you trust a new app

Before you treat a new financial tool like a home base, do one short exercise. Set a timer for 12 minutes and answer these questions in writing (one line each is enough):

1. *Who holds the money?* Name the actual bank/broker/custodian, not the app.
2. *What is the asset?* Deposit? Security? Crypto? Money market fund? Stored value?
3. *What happens if someone steals it?* Is there an error-resolution process? What channel do you use to report?
4. *What happens if the company fails?* Is there deposit insurance, brokerage protection, or nothing comparable?
5. *Where is the adult phone number?* The official support number you would call in a crisis (not a link from a message).

If you can't answer the first two questions quickly, don't connect your paycheck to it, don't store your emergency fund there, and don't let it become your only dashboard. That's not distrust. That's governance.

When the algorithm breaks, you don't win by being the calmest person in the room. You win by being the person who knows which umbrella they're standing under—and exactly what to do when the rain starts.

8.4 The 15-Minute Audit: Evaluating New Financial Tools

The download button is engineered to feel like progress.

You see a clean dashboard, a few animated charts, and a promise that sounds like relief: "Connect your accounts and we'll do the hard part." The hardest part, of course, is exactly what the button is asking for—permission to look inside your financial life, and sometimes permission to act inside it.

Most people treat that moment like installing a flashlight app. They shouldn't. Connecting a new financial tool is closer to giving a stranger a badge that opens doors in your house. Maybe they only enter the kitchen. Maybe they only look around. But once you hand over the badge, you're no longer evaluating a product. You're managing a relationship.

A 15-minute audit is how you keep that relationship from turning into an expensive misunderstanding.

The rule: don't start with features; start with failure

Feature marketing is designed to make you imagine your best self: organized, optimized, calmly wealthy. Due diligence asks a different question: *what happens if this breaks?*

That sounds negative, but it's deeply practical. When a tool breaks, it usually breaks in predictable ways:

- it makes the wrong recommendation confidently,
- it leaks more data than you assumed,
- it pulls money at the wrong time,
- it locks you out and routes you to a chatbot,
- it changes pricing or terms and dares you to leave.

The audit is built to surface those risks before you've built your routines around the tool.

The goal isn't to be suspicious of everything. The goal is to be deliberate about the tools that get to touch the levers.

Minute 1--3: What is this thing, really?

A shocking number of financial products are *two products wearing one face*: a regulated core underneath and a fast-moving tech layer on top. Your first job is to figure out what category you're dealing with. Categories aren't semantics; they determine what you can reasonably expect when things go wrong.

Ask three blunt questions:

1) Does it move money, or only analyze money?

"Read-only" tools can still be risky (data can be weaponized), but the blast radius is smaller than tools that can transfer funds, trade, borrow, or change beneficiaries.

2) Is it acting like an adviser, a broker, a bank, or a media company?

An "AI investing coach" might be giving general education, or it might be steering you toward transactions. The difference matters because obligations and oversight differ. If the tool's output feels like "you should buy *this* now," treat it like advice even if it calls itself "information."

3) Who holds your assets when you use it?

If it's an investing product, where are the securities custodied? If it's "cash," is it a bank deposit, a money market fund, or something else? If it's crypto, who controls the keys?

If you can't answer these in a few minutes from the app's disclosures, that's not a sign you're not smart enough. It's a sign the product may be designed to glide past the questions you should be asking.

Minute 4--6: How does it make money (and what behavior does that reward)?

Money is the loudest signal in the room because it never forgets what it wants.

A tool that charges a flat subscription has one set of incentives: keep you paying, keep you engaged, keep you feeling like canceling would be risky. A tool that makes money on trading volume has another: encourage activity. A tool that routes you into loans or cash advances has another: normalize borrowing.

You don't need a forensic accounting background. You need to be able to say, in one sentence:

"They get paid when I __________."

Fill in the blank:

- trade more,
- deposit more,

- borrow,
- keep idle cash inside their ecosystem,
- share data,
- refer friends,
- upgrade to premium.

Then add one more sentence, the one most people skip:

> "That means the AI will be subtly rewarded for recommending __________."

This isn't accusing anyone of fraud. It's acknowledging gravity. Business models pull behavior in predictable directions, especially when an algorithm is trained to maximize engagement or revenue proxies.

If a tool says it's free, treat your data and attention as the likely product until proven otherwise. "Free" doesn't mean "bad." It means "paid indirectly."

Minute 7--9: What data are you giving it, and for how long?

A modern financial tool isn't just a calculator. It's a vacuum cleaner attached to your life.

When you connect accounts, you may be granting access to:

- transaction history (which reveals habits, health signals, location patterns),
- balances,
- account numbers (sometimes masked, sometimes not),

- employer info via direct deposit patterns,
- recurring bills and subscriptions,
- debt payments and delinquency signals,
- your contact information and device identifiers.

The most important question isn't "Do they have access?" It's *"What can they do with the access?"*

Look specifically for four phrases in privacy notices and data permissions:

- *Share* (with whom? for what purpose?),
- *Sell* (sometimes disguised as "share for marketing"),
- *Affiliates* (this can be broad),
- *Service providers* (normal, but worth understanding).

Then ask the question that separates a temporary experiment from a long-term entanglement:

> If I stop using this, can I revoke access in one minute and be confident it's truly disconnected?

A tool that makes disconnecting difficult is telling you something about its strategy. It wants to become your default. Defaults are powerful. Defaults are also where regret likes to hide.

Minute 10--12: Security and support: can you reach a human when your money is on fire?

Security is rarely the reason people adopt a tool. It becomes the reason they wish they hadn't.

You're looking for two kinds of maturity: account protection and operational support.

Account protection signals.

Good signs include:

- strong multi-factor authentication options,
- device management (seeing and logging out active sessions),
- alerts for logins and profile changes,
- clear recovery paths that don't depend on a single compromised channel.

A tool that asks you to connect accounts but doesn't offer basic security controls is like a luxury apartment with no deadbolt. The marble countertops don't matter.

Operational support signals.

Find the emergency contact path *before* you need it. Is there a phone number? Are hours listed? Is there a secure message center? Do they publish a process for disputes, holds, or errors?

If the only path is "chat with our AI assistant," assume that in a crisis you will be negotiating with a polite wall.

A quick stress test: search within the app for "unauthorized," "fraud," "dispute," "freeze," "lock," "error." If those words are hard to find, that's not an accident. It's a business decision.

Minute 13--15: The credibility check you can do in your pajamas

Most people skip verification because they assume it requires expertise. It doesn't. You just need the right tools.

Verify registrations when advice or brokerage activity is involved.

If a person or firm is acting like a broker or investment adviser, basic verification should be frictionless. A fast way to check backgrounds and registrations (and see disclosures and disciplinary history) is FINRA's BrokerCheck. You're not looking for perfection; you're looking for reality.

Here's the gut-level translation: *If a tool is comfortable taking your money, it should be comfortable telling you who it is.*

Check whether the "team" is a team.

If the company hides the people behind it—no leadership, no physical address, no clear legal entity—treat that as a meaningful risk, especially if they ask for account connections or custody.

An anonymous team might be building a legitimate open-source project. It might also be building a fast exit ramp for itself. Your job isn't to guess which. Your job is to decide what level of risk is appropriate for your money.

Find the exit.

Every serious relationship needs a breakup plan.

Before you connect anything, answer:

- Can I export my data?

- Can I close the account without calling?
- Can I revoke connections from the bank side?
- Can I transfer assets out (if investing)?

A tool that makes exits difficult is telling you it expects dissatisfaction.

A worked example: the irresistible AI trading bot

Imagine a hypothetical app: *AuroraBot*. The ads are clean. The testimonials are confident. The pitch is seductively specific: "Our AI scans thousands of signals and trades for you. Target returns: 2–3% monthly."

That number isn't just a promise. It's a lure. The human brain loves steady, frequent wins. "Monthly" feels manageable, like rent. It also quietly bypasses the question that should hit like a cold glass of water: *If this is so reliable, why are they selling it to me?*

Run AuroraBot through the 15-minute audit:

What is it?

It moves money (requires account connection and trading permissions). That immediately raises the standard. A wrong recommendation is annoying; a wrong trade is real.

How does it make money?

Maybe it's "free" but takes a spread, routes trades, or charges performance fees. If it profits from trading frequency, the AI has a built-in incentive to trade. A model that trades to generate revenue can look brilliant in calm markets and catastrophic in choppy ones.

What data does it want?

It asks for transaction history, balances, and "full access to optimize execution." That phrase is doing heavy lifting. "Optimize" can mean "move fast." "Full access" can mean "you'll find out what it did after it did it."

Security and support?

If AuroraBot has weak MFA and no real-time alerts, you've just placed a high-powered tool in a low-security shed. If the support path is only chat, your ability to stop damage in real time is limited.

Credibility check?

If there's no clear registered entity, no transparent disclosures, and the team is a set of first names with headshots that look like stock photos, the audit doesn't need to prove it's a scam. It only needs to conclude: *this is not a tool to connect to my bank account.*

Here's the subtle lesson: the audit doesn't exist to make you cynical. It exists to make you *hard to rush*. Most bad financial decisions are made under time pressure, excitement pressure, or shame pressure. A 15-minute audit is a small ritual that breaks the spell.

The ``permission ladder'': start low, climb slowly

One of the most effective safety moves is also the least dramatic: don't hand out maximum permissions on day one.

Treat permissions like a ladder:

- Start with read-only access, if possible.

- Then allow small, capped actions (limited transfers, limited trades).
- Only later, if the tool earns it, consider deeper automation.

People often do the reverse because onboarding flows are designed that way. "Connect everything for the best experience." Translation: "Give us more data and more control so we can lock in your habit."

A CEO-minded user flips the default: minimum necessary access, increased only when trust is earned through time and performance.

What ``trust" looks like in practice

Trust is not a feeling you get from a pleasant interface. Trust is what's left after you've asked uncomfortable questions and the answers were clear.

A trustworthy tool tends to have:

- plain-language disclosures that don't require interpretation gymnastics,
- obvious ways to revoke access,
- sensible security defaults,
- a business model that doesn't depend on you doing something frantic,
- support channels that exist outside the app.

And, perhaps most importantly, it behaves well when you're not paying attention. It doesn't punish you for not checking daily. It doesn't turn your anxiety into a revenue stream. It doesn't use AI to manufacture urgency.

Actionable takeaway: the 15-minute audit script

The next time you're tempted by a shiny financial tool, don't fight the temptation. Use it. Just use it as fuel for a fast, repeatable script.

Open a note and write these headings. Answer each in one or two sentences:

1. *Job:* What do I want this tool to do—specifically?
2. *Powers:* Can it move money? Can it trade? Can it borrow in my name?
3. *Pay:* How does it make money, and what behavior does that reward?
4. *Data:* What data does it collect, share, or retain—and can I revoke access quickly?
5. *Security:* What authentication, alerts, and recovery options exist?
6. *Support:* If $5,000 disappears today, what exact number do I call?
7. *Exit:* Can I export, disconnect, and close cleanly?

If any answer feels fuzzy, treat that fuzziness as information. "Unclear" is not neutral. "Unclear" is risk.

And when you want a final decision without overthinking, use the flowchart.

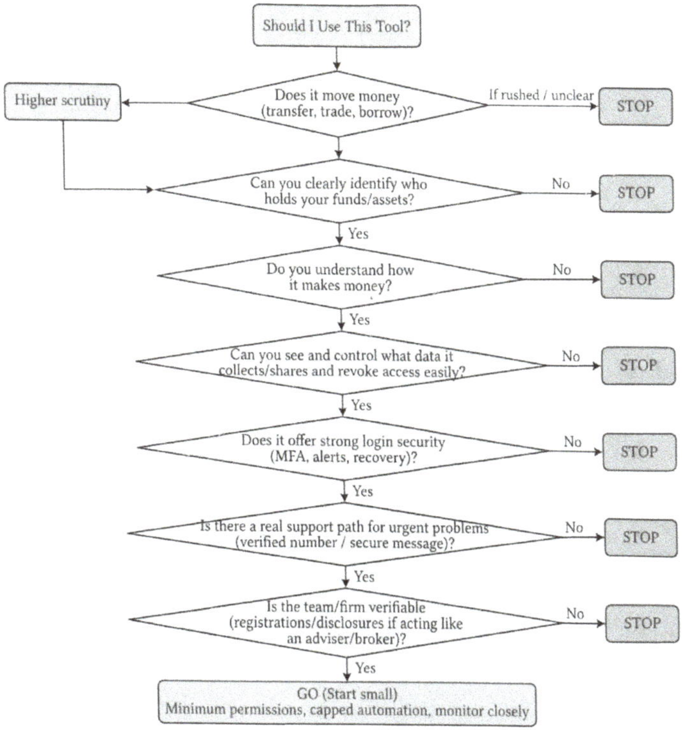

8.5 Trust but Verify: Building a Sustainable Monitoring Routine

Automation has a seductive promise: you do the responsible thing once, and then the responsible thing keeps happening forever.

Auto-save. Auto-invest. Auto-pay. Auto-rebalance. Auto-categorize. Auto-alert.

The words make your financial life sound like a modern dishwasher: load it, press a button, wake up to clean plates. But money doesn't live in a sealed appliance. It lives in the messy ecosystem of your job, your health, your family, your habits, your temptation, your fear—plus a shifting landscape of fees, fraud tactics, and "product updates" that can change what you agreed to without changing the way the app smiles at you.

The danger isn't that automation fails loudly. The danger is that it succeeds at the wrong job quietly.

A sustainable monitoring routine is how you keep the robot working for you without turning your life into a surveillance operation. The goal isn't to hover. The goal is to stay in charge.

A good routine feels boring (and that's how you know it works)

People who are good with money aren't necessarily people who love money. They're often people who love *not thinking about money*—because their systems are reliable enough to fade into the background.

That's the standard to aim for: a monitoring routine that is so lightweight you'll actually do it, and so consistent that it catches problems early.

If a routine requires motivation, it will fail on the exact week you need it most: when you're busy, stressed, traveling, sick, moving, grieving, starting a new job, dealing with a child's crisis, or simply tired of adulthood.

So design the routine like a seatbelt. You don't fasten a seatbelt because you plan to crash. You fasten it because you don't want to make a high-stakes decision at highway speed.

The three drifts: settings drift, incentives drift, and life drift

Most "money problems" caused by algorithms are not sudden catastrophes. They're slow drifts that compound.

Settings drift.

A tool's settings can change because you change them, the tool changes them, or your situation changes and the same setting now means something different. A $500 auto-invest was "aggressive but fine" when you lived with roommates and had low fixed costs. It becomes "quietly reckless" after you take on childcare or a mortgage.

Incentives drift.

The company behind the tool can change its business model. Features get moved behind a paywall. "Free" becomes "freemium." A gentle nudge becomes a persistent upsell. A product starts recommending the option that happens to be best for the company this quarter.

Nothing about your goals changed. The tool's motivations did.

Life drift.

Your financial plan is not a tattoo. It's a living document, even if it only exists as a few sentences in your head. Your risk tolerance changes when your job changes. Your cash needs change when your health changes. Your willingness to "ride out volatility" changes when a down payment becomes real.

This is why the phrase "set it and forget it" is dangerous. The forgetting isn't the problem. The *unchanged automation* is.

What institutions know: risk management is a loop, not a gate

Big organizations don't manage risk by making one smart decision and then congratulating themselves. They manage risk as a cycle: define what you're trying to do, measure what's happening, compare it to the intent, and adjust.

That's true for financial risk, cybersecurity risk, and—now that models and algorithms sit between you and your money—model risk too.

In household terms, "governance" doesn't mean meetings. It means you have recurring moments where you ask:

- Is the system doing what I hired it to do?
- If not, is it a bug, a drift, or a change in me?
- What's the smallest adjustment that gets me back on track?

Those questions don't require obsession. They require a calendar.

The Manager's Schedule: daily glance, monthly sweep, quarterly board meeting

A sustainable routine works because it separates monitoring into different time horizons. Not everything deserves daily attention. Not everything can wait a year.

Here's a rhythm that fits most households without turning you into a full-time compliance officer.

Daily (30--60 seconds): the ``red flag" glance

This is not a budget review. It's a pulse check.

Look for:

- a transaction you don't recognize,
- a balance that looks "off" relative to your normal rhythm,
- an alert about a new login, password reset, or profile change,
- a failed bill payment or overdraft warning.

You're not trying to optimize. You're trying to catch fires while they're still smoke.

The reason this matters in an AI era is that fraud and error chains move quickly. A small, early alert—new device login, contact info change—often comes before the money actually moves. If you only pay attention after money moves, you're already in the hard part of the process.

If daily feels too frequent, do it every other day. The key is consistency.

Monthly (10--20 minutes): the ``systems sweep"

Once a month, you do something that feels oddly satisfying: you check whether your automated systems still make sense.

Focus on:

- Autopays: anything new, anything duplicated, anything that quietly increased.
- Subscriptions: especially the ones that hide behind "annual renewal."
- Automatic transfers: are they pulling too hard (causing tightness) or too soft (letting cash pile up without purpose)?
- Categorization sanity: are your big categories roughly true (housing, food, transport, debt)? Not perfect—true enough.

This is also the moment to check your "money movement hierarchy" in practice. If you've shifted spending to a credit card for safety, are you paying it in full? If you've made checking your buffer zone, is the buffer still there?

Monthly review is where you keep automation from becoming a runaway train. You're not micromanaging. You're keeping the tracks aligned.

Quarterly (30--45 minutes): the ``board meeting"

Quarterly is where you stop thinking like a user and think like the CEO of your financial stack.

Do four things:

1) Reconfirm the plan in plain language.

Write one paragraph answering:

What is money supposed to do for us in the next 12 months?

This isn't a philosophical exercise. It's a calibration. A plan written in plain language is harder for a model—or a salesperson—to hijack with complexity.

2) Recheck risk settings against life reality.

If you use automated investing, revisit the assumptions embedded in your tool settings:

- time horizon,
- emergency fund adequacy,
- major upcoming expenses,
- job stability,
- whether you'd actually tolerate a drawdown right now.

If your life has become more fragile, your portfolio doesn't need a motivational speech. It needs a risk adjustment.

3) Review permissions and connections.

Quarterly is the perfect cadence to prune:

- linked apps,
- data-sharing authorizations,
- devices logged into key accounts,
- third-party access to banks and brokerages.

This is the household equivalent of patching and access reviews. It's not paranoia; it's maintenance. Tools accumulate connections the way kitchens accumulate utensils.

4) Check for ``silent changes."

Look for:

- fee changes,
- new pricing tiers,
- updated terms,
- new default features you didn't request,
- nudges that feel more aggressive than before.

Companies evolve. Your consent should keep pace with that evolution, not lag years behind it.

Monitoring without anxiety: measure the right numbers

A lot of money anxiety comes from staring at the wrong metrics too often.

If you check your investment account daily, you are basically signing up for a tiny emotional rollercoaster subscription. Markets wiggle. Algorithms rebalance. The account value moves. Your brain interprets motion as meaning.

A better approach is to decide which numbers deserve attention at which frequency.

Weekly numbers (for most people).

- Checking buffer balance (are you safe for upcoming autopays?).
- Credit card balance relative to your planned payoff.

Monthly numbers.

- Savings rate (what actually moved into savings/investing).
- Fixed-cost coverage (housing, debt, insurance—are these still manageable?).

Quarterly numbers.

- Net worth trend (direction matters more than precision).
- Asset allocation drift (did your portfolio quietly become something else?).

The objective is not to avoid looking. It's to look at a frequency that matches the timescale of the decision you can actually make. Checking long-term investments daily is like stepping on a scale every hour. You'll get more data and less wisdom.

The ``trust but verify" playbook for AI recommendations

Modern tools don't just execute. They *suggest*. And suggestions can be more dangerous than execution because they feel like your own thoughts.

When an AI tool recommends a move—switch banks, refinance, invest more aggressively, buy a new insurance policy, rebalance, pay down debt instead of investing—run a quick verification loop:

Ask what it assumed.

Every recommendation implies assumptions: income stability, expense stability, time horizon, risk tolerance, tax bracket, fees, and what "success" means. If the tool doesn't make its assumptions visible, you can't evaluate the recommendation. That opacity is model risk.

Ask what it didn't see.

Tools don't see everything:

- a health issue you're managing privately,
- the fact that your job is shaky,
- a family obligation you haven't told anyone about,
- a looming car replacement,
- your true stress level.

A model can only optimize what it can represent. Your real life contains important variables that aren't in the dataset.

Verify via an independent channel.

This isn't about distrusting the tool; it's about reducing single points of failure. Independent channels include:

- direct statements from banks/brokerages (not aggregator views),
- a second calculator (even a simple spreadsheet),
- a human professional for big, irreversible decisions.

That last bullet matters. A tool can be excellent at everyday guidance and still be the wrong instrument for a one-time, high-stakes move like rolling over a retirement account, exercising stock options, or choosing a mortgage product.

Firing the robot: knowing when a tool no longer deserves your trust

People cling to tools for the same reason they cling to bad gym memberships: sunk cost. You've set it up. You've categorized the transactions. You've trained it with your habits. The dashboard feels familiar. It feels like "your system."

But tools don't get loyalty points. They get performance reviews.

A tool should be fired when:

- it repeatedly makes high-impact mistakes and doesn't improve,
- its incentives have drifted into aggressive upselling,
- it becomes harder to reach support as you rely on it more,
- it resists exporting your data or revoking access,
- it asks for broader permissions than its job requires.

The key emotional shift is to treat switching costs as a *tax you pay to remain free*. Freedom is not free. Sometimes it costs you a weekend of admin. The alternative is being trapped in a system that quietly stops serving you.

The stealth risk: automation masking cash-flow fragility

One of the strangest effects of modern automation is that it can make you feel financially stable right up until you're not.

Autopay hides the friction of paying bills. Auto-invest hides the friction of saving. A tool can "optimize" your cash by keeping checking balances low and moving money out aggressively. That can be mathematically efficient and emotionally disastrous if your life includes irregular income, surprise expenses, or any meaningful risk of delayed paychecks.

A monitoring routine protects against this by treating cash-flow stability as a first-class metric. The monthly sweep isn't just about subscriptions; it's about making sure your automations aren't accidentally building a house of cards.

A practical move: choose a checking buffer number that reflects your reality (not your aspiration) and monitor it. If you keep dipping below it, that's not a personal failure. It's a systems signal that either spending, income timing, or automation settings need adjustment.

A routine for the household, not just the individual

In many households, one person becomes the CFO by default. Sometimes they like it. Often they don't.

The monitoring routine becomes more sustainable when it's shared, even lightly:

- A monthly 10-minute "money sweep" together, even if one person does most of the clicking.
- Shared rules for verification when messages are urgent (call-back habits, safe word).
- Agreement on thresholds: "Anything above $X needs a quick check-in before we move it."

This reduces risk in two ways. First, it reduces the single point of failure that comes from one person holding all the knowledge. Second, it makes fraud harder. Social engineering often targets the person least likely to verify. Shared routines raise the household's baseline skepticism in a healthy way.

Actionable takeaway: the one-page monitoring checklist you'll actually use

Open your calendar right now and schedule three recurring events. Name them exactly like this so Future You understands what Past You intended.

1. *"Money Red Flags"* (every 2–3 days, 1 minute)
   Check alerts, unknown transactions, unusual balances. No optimization. Just smoke detection.
2. *"Systems Sweep"* (monthly, 15 minutes)
   Review autopay, subscriptions, transfer rules, and checking buffer. Adjust one thing if needed.
3. *"Quarterly Board Meeting"* (every 3 months, 45 minutes)
   Reconfirm goals in plain language, review risk settings, prune permissions, and scan for silent fee/term changes.

Then add one sentence to the top of your notes app and pin it:

> Automation is allowed to run only if it still matches our life.

That sentence is the whole philosophy. You can trust tools to do what they're designed to do. You just can't trust them to notice when your life changes, when incentives change, or when a model starts confidently walking toward the wrong cliff. Your routine is the human loop that keeps the system pointed in the direction you actually care about.

www.ingramcontent.com/pod-product-compliance
Lightning Source LLC
LaVergne TN
LVHW052028200926
848481LV00005B/1617